**C
O
N
N
E
C
T
E
D**

GET

Tech
Programs
for Teens

**RoseMary
Honnold**

*for
the Young Adult
Library Services
Association*

BIBLIOTHÈQUES

u Ottawa

LIBRARIES

Neal-Schuman Publishers, Inc.

New York   London

Published by Neal-Schuman Publishers, Inc.
100 William St., Suite 2004
New York, NY 10038

Printed and bound in the United States of America.

The paper used in this publication meets the minimum requirements of American National Standard for Information Sciences - Permanence of Paper for Printed Library Materials, ANSI Z39.48-1992.

ISBN-10: 1-55570-613-4
ISBN-13: 978-1-55570-613-5

Library of Congress Cataloging-in-Publication Data

Honnold, RoseMary, 1954–
    Get connected : tech programs for teens / RoseMary Honnold for the Young Adult Library Services Association.
       p. cm.
    Includes index.
    ISBN 978-1-55570-613-5 (alk. paper)
1. Libraries and teenagers—United States. 2. Young adults' libraries—Activity programs—United States. 3. Internet in young adults' libraries—United States. 4. Internet and teenagers—United States. 5. Technology and youth—United States. 6. Information literacy—Study and teaching (Secondary)—Activity programs—United States. I. Young Adult Library Services Association. II. Title.
    Z718.5.H66 2007
    027.62'60973—dc22

                                                          2007012847

Contents

List of Figures

Contributors

Thank you to the generous contributors to *Get Connected: Tech Programs for Teens:*

Alessio, Amy
Teen Coordinator
"Gamers' Group"
aalessio@stdl.org
Schaumburg Twp. Dist. Library
130 S. Roselle Rd.
Schaumburg, IL 60193
(847) 923-3191
www.stdl.org
www.myspace.com/stdlteen

Angell, Michelle
Youth Services Librarian, Graham
    Branch
"Meetings and Myspace: Mylibrary
    Teen Council"
mangell@pcl.lib.wa.us
Pierce County Library System
    Graham Branch
PO Box 1267
9202 224th St.
East Graham, WA 98338
(253) 847-4030
www.pcl.lib.wa.us/

Arnold, Mary
Teen Services Manager
"Pass the Book"
marnold@cuyahogalibrary.org
Cuyahoga Public Library
2111 Snow Rd.

Parma, OH
(216) 475-5000
www.cuyahogalibrary.org

Carson, Grier
Librarian
"MP3s at LFA"
gcarson@lfanet.org
Lake Forest Academy Library
1500 W. Kennedy Rd.
Lake Forest, IL 60045
(847) 615-3205
www.lfanet.org/curriculum/library.htm

Coffey, Clare
Senior Librarian, Humanities
    Department
"Teen Tech Camp"
coffeyc@memphislibrary.org
Memphis Public Library and
    Information Center
3030 Poplar Ave.
Memphis, TN 38111
(901) 415-2700
www.memphislibrary.org
http://memphisreads.blogspot.com/
http://memphismusic.wordpress.com

Conelly, Megan
Library Assistant II
"Teen Video Game Tournament"

meconelly@washoecounty.us
Mendive Community Library,
  Washoe County Library System
1900 Whitewood Dr.
Sparks, NV 89434
(775) 353-5989
www.washoe.lib.nv.us/

Czarnecki, Kelly
Teen Librarian
"*Soul Calibur II* Tournament"
"Be Smart-Wired"
kczarnecki@plcmc.org
ImaginOn: The Joe and Joan Martin
  Center, Public Library of Charlotte
  & Mecklenburg County
300 East 7th St.
Charlotte, NC 28202
(704) 973-2728
www.imaginon.org
www.myspace.com/libraryloft

Declet, Jaime
Children's Librarian
"*Stepmania*"
JDeclet@LPLS.info
Lorain Public Library System
351 Sixth St.
Lorain, OH 44052
(440) 244-1192
www.lorain.lib.oh.us

Demarchi, Bonnie
Teen Services Manager
"Unplugged @ Your Library—READ"
Bdemarchi@Cuyahogalibrary.org
Cuyahoga County Public Library
Parma-South Branch
7335 Ridge Rd.
Parma, OH 44129
(440) 885-5362
www.cuyahogalibrary.org

Erwin, Sarah
Manager of Children's & YA Services

"LAN Party"
serwin@real.more.net
Kirkwood Public Library
140 E. Jefferson Ave.
Kirkwood, MO 63122
(314) 821-5770 ext. 15
www.kpl.lib.mo.us

Goldsmith, Francisca
Library Services Manager
"Earphone English"
frg1@ci.berkeley.ca.us
Berkeley Public Library
2090 Kittredge St.
Berkeley, CA 94704
(510) 981-6139
www.berkeleypubliclibrary.org
http://berkeleypubliclibrary.org/
  weblog/teens/

Greenfield, Kelly
Library Information Specialist II
"Teens WAVE @ Central"
kgreenfi@vbgov.com
Virginia Beach Central Library
4100 Virginia Beach Blvd.
Virginia Beach, VA 23452
(757) 385-0150
www.vbgov.com/dept/library/
www.blogspot.cltc.com

Gwin, Candice
IT/Marketing Manager
"LAN Party"
cgwin@real.more.net
Kirkwood Public Library
140 E. Jefferson Ave.
Kirkwood, MO 63122
(314) 821-5770 ext. 15
www.kpl.lib.mo.us

Hess, Sandy
Business Marketing Careers Teacher
"Classroom Performance System"
cj_11@omalp1.omeresa.net

Coshocton County Career Center
23640 Airport Rd.
Coshocton, OH 43812
(740) 622-0211 ext. 118
www.coshocton-jvs.K12.oh.us/

Iser, Stephanie
Children's Librarian
Teen Tech Week Committee
"Computer Block Party"
"Media Space"
"Teen Pen-Pal Podcast"
stephanieiser@kclibrary.org
Kansas City Public Library—L.H.
   Bluford
3050 Prospect Ave.
Kansas City, MO 64128
(816) 701-3689
www.kclibrary.org

Jepson, Ed
Collaborating Head Librarian
"Back to the Future"
bsjh_jepson@omalp1.omeresa.net
Bellaire St. John Central HS Library
3625 Guernsey St.
Bellaire, OH 43906
(740) 676-4932
www.bellairestjohn.sbd.pvt.k12.oh.us/

Kniesner, John
Collaborating Head Librarian
"Back to the Future"
kniesnjo@oplin.org
Bellaire Public Library
330 32nd St.
Bellaire, OH 43906
(740) 676-9421
www.bellaire.lib.oh.us/

Lauzon, Alissa
YA/Reference Librarian
"Haverhill Public Library Teen
   MySpace Page"
alauzon@mvlc.org

Haverhill Public Library
99 Main St.
Haverhill, MA 01830
(978) 373-1586
www.haverhillpl.org
www.myspace.com/haverhill_public_
   library

Lloyd, John
Librarian I, Sciences Department
"Teen Tech Camp"
lloydj@memphislibrary.org
Memphis Public Library and
   Information Center
3030 Poplar Ave.
Memphis, TN 38111
(901) 415-2700
www.memphislibrary.org
http://memphisreads.blogspot.com/
http://memphismusic.wordpress.com/

Makens, Katherine
Young Adult Services Librarian
"Tech Lab"
kmakens@browardlibrary.org
North Regional/BCC Library
1100 Coconut Creek Blvd.
Coconut Creek, FL 33066
(954) 201- 2618
Fax: (954) 201- 2650

Morgan, Sarah Kline
Teen Librarian
"The Cheshire Public Library
   Podcast"
smorgan@cheshirelibrary.org
Cheshire Public Library
104 Main St.
Cheshire, CT 06410
(203) 272-2245
www.cheshirelibrary.org
cpltbb.wordpress.com

Mori, Maryann
Teen Services Librarian

"Tech Teens"
maryann@evpl.org
Evansville Vanderburgh Public
 Library
200 SE Martin Luther King Jr. Blvd.
Evansville, IN 47713
(812) 428-8229
www.evpl.org

Mormon, Ruth
Librarian
"Find eLove in the Library"
rmormon@themeadowsschool.org
The Meadows School Upper/Middle
 School Library
8601 Scholar La.
Las Vegas, NV 89128
(702) 254-1610
www.themeadowsschool.org

Mulligan, Christy
Teen Central Librarian
"Gaming Studio"
clmulligan@mplib.org
Minneapolis Public Library—Teen
 Central
300 Nicollet Mall
Minneapolis, MN 55401
(612) 630-6000
www.mplib.org
www.myspace.com/mplsteencentral

Paddock, Susan
Library Information Specialist II
"Teens WAVE @ Central"
spaddock@vbgov.com
Virginia Beach Central Library
4100 Virginia Beach Blvd.
Virginia Beach, VA 23452
(757) 385-0150
www.vbgov.com/dept/library/
www.blogspot.cltc.com

Paone, Kimberly
Supervisor, Adult/Teen Services

"A Week of Japanese Entertainment"
KPaone@elizpl.or
Elizabeth Public Library
11 South Broad St.
Elizabeth, NJ 07202
(908) 354-6060 ext. 7235
www.elizpl.org

Parks, Angela
Library Associate & YA
 Programming
"*Dance Dance Revolution* @ Your
 Library"
aparks@olatheks.org
Olathe Public Library
201 E. Park St.
Olathe, KS 66061
(913) 971-6881
www.oltathelibrary.org

Politi, Nicole
Librarian, Young Adult Services
"*Madden 2007* Xbox 360
 Tournament"
Politi_N@oceancounty.lib.nj.us
Ocean County Library, Toms River
 Branch
101 Washington St.
Toms River, NJ 08753
(732) 349-6200
www.oceancountylibrary.org
http://tabspace.wordpress.com
http://nicolepoliti.wordpress.com

Rauseo, Melissa
Young Adult Librarian
"YA Fantasy Baseball League"
rauseo@noblenet.org
Peabody Institute Library
82 Main St.
Peabody, MA 01960
(978) 531-0100
www.peabodylibrary.org
www.myspace.com/wallythe
 yawizard

Roach, Matt
Teen Librarian
"*Soul Calibur II* Tournament"
mroach@plcmc.org
Steele Creek Branch Library
  (Public Library of Charlotte &
  Mecklenburg County)
13620 Steele Creek Rd.
Charlotte, NC 28273
(704) 588-4345

Swarzwalder, Jami
Teen Tech Week Committee
jamischwarzwalder@gmail.com

Shannon, John
Assistant Manager/Youth Services
"*RuneScape*"
JSHANNON@columbuslibrary.org
Columbus Metropolitan Library
South High Branch
3540 S. High St.
Columbus, OH 43207
(614) 479-3364
www.teens-connect.com
http://profile.myspace.com/index.
  cfm?fuseaction=user.viewprofile&
  friendid=81235044

Silchuk-Ashcraft, Daphne
Young Adult Librarian
"Homework Helpers"
Silchuk@mcdl.info
Medina County District Library
3800 Stonegate Dr.
Medina, OH 44256
(330) 725-0588
www.mcdl.info/

Snow, Beth
Reference/Young Adult Associate
"Video Production Workshop for Teens"
bethsno@mail.sgcl.org
Library Center, Springfield-Greene
  County Library District

4653 South Campbell Ave.
Springfield, MO 65810
(417) 874-8110 ext. 144
http://thelibrary.org/
http://thelibrary.org/teens/teens.cfm

Spence, Rebecca
Teen & Information Services
"Game On"
rspence@pls.lib.ok.us
Norman Public Library of the
  Pioneer Library System
225 Webster Ave.
Norman, OK 73069
(405) 701-2681
www.justsoyouknow.us/teens
http://teen_initiative.livejournal.com

Tyle, Alexandra
Reference Librarian
"Teen Techies"
"A Digital Imagination: A Digital
  Graphic Art Contest"
atyle123@yahoo.com
Homer Township Public Library
Homer Glen, IL 60491
(708) 301-7908
www.homerlibrary.org/teens.asp
www.homerlibrary.org/teenevents.asp
www.homerlibrary.org/teenreviews.asp
http://librarybookdiscussion.
  blogspot.com/
http://del.icio.us/homrteens

Uhler, Linda
Teen Services Associate
"Teen Night at the Movies"
"Teens Are Good to Go with
  Playaway Audio"
luhler@holmeslib.org
Holmes County District Public Library
3102 Glen Dr.
Millersburg, OH 44654
(330) 674-5972
www.holmes.lib.oh.us

Vieau, Jesse
Senior Library Assistant
Soul Calibur II Tournament"
jvieau@plcmc.org
ImaginOn: The Joe and Joan
 Martin Center, Public Library of
 Charlotte & Mecklenburg
 County
300 East 7th St.
Charlotte, NC 28202
(704) 973-2728
www.imaginon.org
www.plcmc.org
www.libraryloft.org
http://thegamingzone.blogspot.com
www.myspace.com/libraryloft

Whitworth, Kristen
Teen Librarian

"Business Card Workshop"
kwhitworth@greenville.org
Greenville County Library System
25 Heritage Green Pl.
Greenville, SC 29601
(864) 242-5000 ext. 2247
www.greenvillelibrary.org/teenville/
 index.html

Wilson, Betty Anne
Assistant Director for Library
 Advancement
"Teen Tech Camp"
wilsonba@memphislibrary.org
Memphis Public Library &
 Information Center
3030 Poplar Ave.
Memphis, TN 38111
(901) 415-2847

Preface

Today's teens are digital natives. They carry cell phones, rely on the Internet for homework help, and are always forging ahead, ready and eager to experiment with the latest electronics. They listen to music while updating their MySpace profiles, instant messaging, watching videos on YouTube, and searching for more. I frequently stop students in the library to ask about their new gadgets—increasingly smaller tools that do increasingly larger tasks.

Teens often have little guidance in how to use technology. Many ignore owners' manuals in favor of "clicking to see what happens." Their sense of adventure challenges librarians to keep up with what students are doing and guide them towards safe and responsible use. As new technologies expand the scope of information literacy, teaching teens to evaluate information is more critical than ever.

One of the best ways to foster information literacy is by offering programs that appeal to teens' interest in technology. *Get Connected: Tech Programs for Teens* includes educational and recreational programs that will attract teens to the library. Examples in this book reflect the experiences of real libraries and real students. Readers can follow the models closely or adapt them to fit their library's unique situation.

Get Connected also outlines ways to improve your library's Web presence using technologies teens like and use themselves. This makes the library more attractive (and indeed, less alien) to their world. Library 2.0 initiatives focus on involving patrons in the creation and modification of services. Young adult librarians have done this for years through teen advisory groups, but we now have the opportunity to expand the two-way communication electronically. Libraries can use Web 2.0 concepts like blogs, Flickr, MySpace, podcasts, and RSS to provide ongoing information to teens and to receive input from them about the services and materials they need and want. Many institutions have developed helpful Web sites that guide teens to valid resources and offer online safety tips. Others keep teens informed of upcoming events through a library blog and solicit teen opinions through library MySpace pages. Some use screencasts and podcasts for instruction or even promote their services with videos hosted at YouTube.com.

In December 2005, Education|Evolving, a research and advocacy group supported by the Center for Policy Studies and Hamline University, presented a major report, "Listening to Student Voices—On Technology," that provided a snapshot of teens' relationship to technology. The full report is available at: www.educationevolving.org/studentvoices/pdf/tech_savy_students.pdf. Here are just a few of their conclusions:

1. Technology is important to students in education
2. Students want to use technology to learn, and in a variety of ways
3. Students want challenging, technologically oriented instructional activities
4. Students want adults to move beyond using the "Internet for Internet's sake"
5. Students want to learn the basics, too

Too often, we assume that teens are only interested technology because of its entertainment value. The Education|Evolving study reveals that teens also have a strong desire for information literacy. The ideas in *Get Connected: Technology Programs for Teens* will help you provide fun, interesting services that also teach teens how to be savvy technology users.

Finding the Right Program

Each chapter of *Get Connected* describes several creative, successful, tech-based programs. Throughout the book, "Tech Notes" introduce you to new terms, products, and concepts and lists of print and Web resources provide a starting point for further learning.

Part I, "Get Connected for Fun," focuses on recreational programs: game and dance tournaments, podcasts, art and film contests, and tutorials on social networking sites like MySpace. These events attract teens and encourage them to become more involved with the library.

Part II, "Get Connected for Education," offers opportunities for partnerships between schools and public libraries. Instructional programs can teach teens to use their favorite technologies more effectively, do their schoolwork more efficiently, and address challenges such as online job hunts.

Part III, "Get Connected with the Teen Advisory Group," shares ideas for establishing tech-focused teen advisory groups, which can help librarians decide what to teach, what to buy, and what to offer as entertainment. Teen participation in all aspects of young adult services leads to better results.

Appendix A reproduces the questionnaire filled out by the *Get Connected* contributors. The other appendices include the ALA's positions on the importance of technology in libraries and access to patrons of all ages, the YALSA competencies for librarians serving youth, and YALSA's social networking toolkit, created to accompany Teen Tech Week.

All the programs here are ideal ways to celebrate YALSA's Teen Tech Week. The first annual Teen Tech Week, in March of 2007, marked YALSA's 50th anniversary year. No event could have been more appropriate. Our wired and connected teens propel us into the future by introducing us to new ways of finding and delivering entertainment and information.

Ready or not, technology is a significant part of teens' lives and, therefore, your life as a librarian who serves them. The program ideas in *Get Connected* will help you understand and connect with teens and technology. You might even have some fun at the same time!

Part I

Get Connected for Fun

Chapter 1

The Game Connection

Gaming is a significant part of many teens' lives. The Pew study, "Teens and Technology: Youth Are Leading the Transition to a Fully Wired and Mobile Nation," discovered 81 percent of the teens online played games. Video games of all genres have been created to appeal to every audience and all ages. Recent active video games, such as *Dance Dance Revolution* and *Guitar Hero*, get the teens up and moving, and the newest game systems have controllers that take us a step closer to a virtual reality experience. Online role-playing games often appeal to fantasy and adventure readers and vice versa, making a direct connection with your readers and technology. Several online multiplayer game sites are free, requiring no investment in equipment beyond the Internet computers your library already owns. Survey, poll, or just chat with your teens to see what games are popular in your area. They will know the local game store folks and may already have the games and equipment you need to put together a gaming event.

A LAN Party
Afterhours on Saturday nights at the Kirkwood Public Library in Kirkwood, Missouri, is LAN Party time for eighth through twelfth graders. As of this writing, two LAN parties have been held for four hours each, 6 pm to 10 pm. The Children's Room computers and eight laptop computers are networked for the games, and *Battlefield 1942* is the current chosen party game.

Released in 2002, *Battlefield 1942* has become a classic. A first-person shooter game set in World War II, it can be played alone or as a multiplayer game. Players can fly fighters and bombers, navigate ships, drive tanks, or fight in the infantry on maps of famous battlefields. The official site is www.ea.com/official/battlefield/1942/us/

Figure 1.1: LAN Party Group Photo

The Kirkwood Public Library LANners pose for a photo after their *Battlefield 1942* LAN party. (Reprinted with permission of the Kirkwood Public Library.)

The Teen Advisory Board suggested this program and the Children's Room Manager and the IT/Marketing Manager purchased the software. Teens helped install many of the games.

On the evening of the program, teens arrive at 6 pm and immediately begin playing the game. One of the computers is designated as the server. More teens show up than there are computers so they take turns, playing the role-playing fantasy game *Dungeons and Dragons* while waiting for computer time. Pizza arrives at 7 pm, and teens purchase their beverages from library vending machines.

Both the computer games and pizza are purchased using library teen programming money. Domino's gives the library a discount for the pizza. The teens really enjoy playing and are always asking if they can do it again, so more LAN parties are scheduled.

The library is currently searching for funding for upgraded laptops because the game plays better on computers with higher processing capacity.

Submitted by Sarah Erwin and Candice Gwin

✍ What Is a LAN Party? ✍

A LAN party is a group of people who play multiplayer computer games on networked computers in one location. Players usually bring their own computers to a LAN party but a computer lab can be used. Check out http://en.wikipedia.org/wiki/LAN_party for a complete equipment list. Dedicated LAN partiers prefer the term "LANners." LANners prepare for the long haul with energy drinks and large amounts of fast-food nourishment.

● ● ●

RuneScape: Tips, Guides, and Games

The South High Branch of the Columbus Metropolitan Library in Columbus, Ohio, hosted its first RuneScape program for 16 teens, ages 13 to 18, on a September Saturday from 5 pm to 8 pm. Two of the library's most experienced teen RuneScape players designed the program. They suggested tips to share and freebies players would want. The program was promoted through word of mouth.

RuneScape is one of the most popular online games, with millions of players worldwide. It is a quest game set in the fantasy realm of Gielinor, and players can use magic or mechanical means

---

Helpful *RuneScape* Web Sites

Gower, Andrew and Jagex, Ltd. "Rune HQ." Available at: www.runehq.com

"Rune Tips." Available at: http://tip.it/runescape

Rippel, Chris. "Library RuneScape Teams." Available at: www.ckls.org/~crippel/runescape/teams.html

Rippel, Chris. "RuneScape Tournaments." Available at: www.ckls.org/~crippel/runescape/RunescapeTournaments.doc

to travel and encounter monsters and other players. Players can chat, trade, and play games. *RuneScape* can be played through any browser and the Internet. The official *RuneScape* Web site is www.runescape.com.

The first hour of the program was in the meeting room, with pizza, chips, and pop. A laptop, projector, and MS PowerPoint program were used to present tips and tricks to play *RuneScape* and to explain the use of passwords, "friends" and "ignore" lists, chat, bots, macros, and autominers. The teens were eager to share their tips with each other. A map of the *RuneScape* wilderness was distributed, which many teens had said would be great to have. The teens chose the South High Clan's name, Berserker's Fury, and a red cape color.

Once the branch closed at 6 pm, the program moved to the library's public Internet computers. The first task for the South High Clan was to find the librarian in the wilderness and PK, or Player Kill, the character. During this part of the program, each player was added to the librarian's friends list, and questions were answered using chat. Then they began playing *RuneScape* until 15 minutes before the program ended at 8:00 pm.

Figure 1.2: *RuneScape* **PowerPoint Front Page**

The *RuneScape* Tips, Guides and More PowerPoint program was presented while teens snacked on pizza and chips. (Reprinted with permission of the Columbus Metropolitan Library.)

The players enjoyed being allowed to yell out to each other and to run from computer to computer. A number of teens then checked out many of the books displayed for the program, including *Dungeons and Dragons* manuals and fantasy fiction on their way out the door. One of the teens created a message board for the group to post more gaming events.

Figure 1.3: *RuneScape* Tips, Guides, and More!

Rules
- Exchanging RuneScape items for other real-life benefits is not allowed
- You must not attempt to use other programs (e.g., bots, macros, or autominers) to give yourself an unfair advantage at the game

Security
- Don't tell ANYONE your password, ever!
- Don't tell others your password-recovery questions
- Don't tell others your real name, email, ICQ number, credit-card details, etc.

Scamming Prevention Guide
- Be wary of newly created/low level players
- Don't fall for "unofficial" trades!
- Be wary of displays of trust
- Be wary of players offering to make you items

RuneScape Movies
- Download and install a screen recorder, such as HyperCam
- Record the event
- Save to your desktop
- Upload your movie to sites such as YouTube or Google Video

Other RuneScape Clans
- Shadow Legions
- Level 80+ Combat Level *or*
- Level 85+ Magic *or*
- Level 85+ Range *or*
- Princess Mom Clan
- Mom
- Have a princess dress

South High Clan
- Rules
- No flaming
- No attacking other clan members on PK trips
- Respect the leaders
- Name?
- Cape color?
- Message Board?
- PKing John
- Username is "Wakayamadan"
- First world I will visit is 102
- You can keep what you take
- Special prizes

"Rules, Tips, and Tricks for *RuneScape*" was presented in a Power Point at the beginning of the program.

Submitted by John Shannon

✍ What Is a MMORPG? ✍

MMORPG stands for Massively Multiplayer Online Role Playing Game. Large numbers of players can log in and play and roam around a 2D or 3D virtual world, interacting with each other. *World of Warcraft* at www.worldofwarcraft.com/bc-splash.htm is the MMORPG with the most active memberships, but there are many other games. Check the list at: http://en.wikipedia.org/wiki/List_of_MMORPGs

✍ Tech Note ✍

Get connected with other libraries offering gaming programs. Sign up for LibGaming, a discussion group dedicated to gaming in libraries, at: http://groups.google.com/group/LibGaming

● ● ●

Stepmania Dance Competition

Teens in grades six through twelve gather once a month after school for two hours to dance the afternoon away in the meeting rooms of Lorain Public Library in Lorain, Ohio. "*Stepmania:* Join the Dance Revolution!" was offered to see if there was a teen audience to possibly develop a Teen Advisory Group.

Video games have taken on a new attitude with physical challenges that exercise the body while exercising the mind, coordination, and reflexes. Dancing programs are popular in arcades and now in libraries. *Stepmania* is a low-cost, open-source variation of the *Dance Dance Revolution* type program that will help you decide if a dance program works in your library. *Stepmania* works for Windows, Mac, and Linux and features 3D graphics, dance pad support, and a Step Editor for creating files of dance moves.

Dance pads and refreshments are the only expenses with the *Stepmania* program. Four pads and four jacks that allow connection between the pads to a PC were purchased for under $40. The computers this library used were computers stored in a closet to use for spare parts from the technology department!

Gaming magazines and gaming books were on display during the programs. The programs were meant to last until 6 pm, but teens were still having fun at 8 pm. They signed up right away

Figure 1.4: *Stepmania* Publicity Flyer

Join the revolution, the dance revolution.

at the Lorain Public Library System's
Main Library, 351 Sixth Street, Lorain
Show off your moves at this fast-paced video dancing game
party complete with refreshments and snacks.
For sixth through twelfth graders.
For more information, call the Lorain Public Library System's Main
Library at 440-244-1192 or 1-800-322-READ, ext. 246.

Wednesdays at 4 p.m.
September 20
October 18
November 15

LORAIN
PUBLIC LIBRARY
S Y S T E M
www.lorain.lib.oh.us
Where Learning Is For Life

A publicity flyer helped spread the word about Lorain Public Library's *Stepmania* dance program. (Reprinted with permission of the Lorain Public Library.)

for the next dance program. If *Stepmania* continues to be a success, the library plans to apply for a grant to buy several Nintendo Game Cubes and Playstation consoles, the *Dance Dance Revolution* software, *Guitar Hero,* and a number of other games to implement a steady gaming program at the library. The library is also working towards getting the local video store involved with the programs by either bringing equipment or even running quarterly tournament events.

Stepmania is available as a free download at www.stepmania. com. The Web site has many links to music downloads, information about dance pads and controllers, and a photo gallery of *Stepmania* events in libraries and schools in Europe and the United States. The "Recommended" link lists advantages and disadvantages of the various types of dance pads.

Submitted by Jaime Declet

● ● ●

Soul Calibur II Tournament

A *Soul Calibur II* tournament was held over the summer at eight branches of the Public Library of Charlotte and Mecklenburg County in Charlotte, North Carolina. Teens had suggested a game tournament and several games for it; *Soul Calibur II* was their final choice. Eight weekly two-hour qualifying rounds were promoted with a video created by teens at the library's Web site, www.libraryloft.org, where two qualifying players' names from each branch were posted after each round.

Soul Calibur II is the updated version of *Soul Calibur*, a fighting game with knights and swordplay set in Europe in 1591. Soul Calibur is the name of the holy sword created to battle the evil sword, Soul Edge. *Soul Calibur II* comes in versions to play on Xbox, Nintendo GameCube, and Playstation 2. The Web site is www.soulcalibur.com.

Teens ages 12 to 18 were invited to each event. Three Xbox consoles were available to play *Soul Calibur II*. The library collaborated with GameLAN'd (www.gamelandonline.com), a local shop where one can pay $5 per hour to play any game on one of their many flat-screen monitors. GameLAN'd donated a few all-day passes to their establishment to the top players in the tournament. Many different books were on display, including gaming strategy guides, graphic novels, and the art of producing video games.

Each of the eight locations had at least one teen librarian running the tournament. Most of the locations had co-workers helping them, but the event can be held with just one supervising person. Training sessions were held at participating library branches on how the tournament would run at each location.

Each of the eight qualifying rounds had two champions who received a $10 gift certificate to EB Games. In the championship round, the runner-up received another $10 gift certificate and the grand prize winner received a $50 gift certificate, both to EB Games. Total cost for all prizes was $230. Other expenses may include additional gaming consoles and games.

The following schedule was used to set up the tournament:

- Eleven weeks prior to program: Choose a game format (Xbox, Playstation 2, GameCube, PlayStation, Sega, any of the Nintendo consoles, etc.)
- Ten weeks prior to program: Create a core team of co-workers interested in setting up a county-wide tournament. Discuss all tournament rules, including elimination standard (double/single), program time limit, participant limits (number, age), warm-up round time limit, allowing same players at more than one qualifying round (at eight separate libraries), numbers of winners per branch, and number and types of prizes (e.g., gift cards from local gaming merchants)
- Eight weeks prior to program: Send e-mails to branches to see which ones will participate
- Seven weeks prior to program: Visit branches to determine equipment needs
- Six weeks prior to program: Create promotional flyers, videos
- Three weeks prior to program: Host training session for participating branches
- Two weeks prior to program: Follow up on any questions from training session
- One week prior to program: Create evaluation forms, contact all participating branches to be sure they are ready to host their qualifying round, advertise at all other teen programs, send out reminders to e-mail lists
- Gameday: Allow a 30-minute warm-up round and sign each participant in as they enter the room. After 15 to 20 minutes of warm-ups, allow no one else to join. Make the bracket by using the sign-in sheet. (If doing a double elimination tournament: Go to the tournament bracket Web site at: www.crowsdarts.com/brackets/tourn.html and print out the correct bracket for the number of participants.)

Tournament Evaluation

What program did you attend? _____

How did you hear about this program?

Flyer in the library   Program or poster in the library   Friend or parent
Teacher at school   Library Web page   E-mail
Newspaper   Other

How did today's program rate with you?

Great   Very Good   Average   Poor

What other games would you be interested in playing in a tournament-style event?

Any suggestions to make this program better?

What other types of programs would you be interested in attending at the public library?

What is the best time for you to attend programs? (after school, nights, or weekends?)

Do you have any general comments about the library and/or its programs for teens?

Have you ever used the library's Web page? [www.libraryloft.org]
( ) yes   ( ) no

ImaginOn's MySpace? [www.myspace.com/libraryloft]   ( ) yes   ( ) no

OPTIONAL: May we contact you to gather more information?
( ) yes   ( ) no

Your name, e-mail address, and telephone number are private. It is available only to appropriate library staff in support of library service.

Name
Phone
E-mail

The teens all loved the tournament and immediately asked when the next tournament would be held. They had many suggestions on what to play next and how the program might run better in the future. One even created his own MySpace page dedicated to being the "*Soul Calibur II* champion of Mecklenburg County." Future plans include building the championship round into a much bigger event and finding some celebrity players to attend and compete against the teens. The tournaments will be advertised on TV and radio and more community collaboration will be pursued.

Figure 1.5: Library Loft Logo

The Library Loft logo brands teen space and activities at the Public Library of Charlotte and Mecklenburg County. (Reprinted with permission of the Public Library of Charlotte and Mecklenburg County.)

Figure 1.6: *Soul Calibur II* Program Flyer

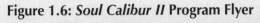

Teens could decide when and where to qualify for the *Soul Calibur II* Tournament with the information on the flyer. (Reprinted with permission of the Public Library of Charlotte and Mecklenburg County.)

Figure 1.7: *Soul Calibur II* Web Site Screen Shot

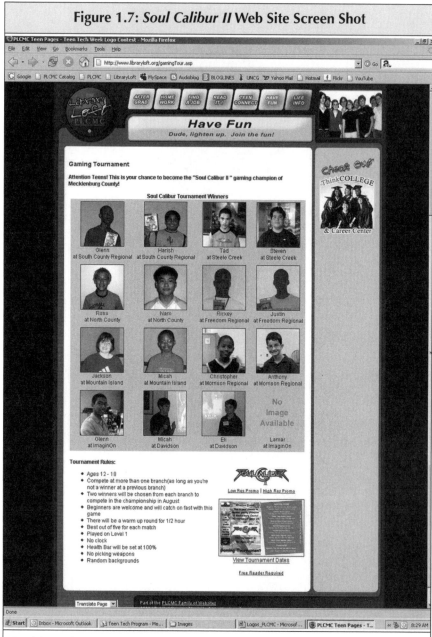

Qualifying players were posted on the Web site after each qualifying round. (Reprinted with permission of the Public Library of Charlotte and Mecklenburg County.)

Submitted by Jesse Vieau, Kelly Czarnecki, and Matt Roach

● ● ●

Gaming Extravaganza

After a successful initial gaming program in the summer at Norman Public Library in Norman, Oklahoma, teens begged for a monthly event. Teens, grades seven through twelve, now come to the library meeting room for the "Game On!" gaming extravaganza every month. Xbox, PlayStation 2, Game Cube, and *Dance Dance Revolution* consoles are set up by two staff members and five teen volunteers. GameStop gave the library a 20 percent discount when they purchased the game systems with funds provided by the Friends of the Library. The total investment was about $1,000. The events are promoted at the local school libraries.

On game day, setup for the event begins at 8:00 am. There are stations for eight games: two stations of *Dance Dance Revolution* and one of *Madden 06, Smash Brothers, Street Football, Call of Duty, Gauntlet,* and *Mario Cart.* Board games are set out also: *Buffy the Vampire Slayer, Star Wars Game of Life, Millennium Trivial Pursuit, Star Wars Monopoly,* and *Fact or Crap.* Doors open to the teens at 10:00 am. Players stream in slowly until about noon, when the room completely fills up. Bottled water and popcorn are served at noon. The crowd begins to dwindle about 3:00 pm. Pack-up begins at 3:30 pm and the doors close at 4:00 pm. Game packing, room cleaning, and such is completed by 5:00 pm.

The *Street* sports games seem to be much more popular than programs like *Madden 06.* The biggest attractions are the *Dance Dance Revolution* and the *Smash Brothers* games. The library is looking at purchasing more interactive equipment such as drum and guitar games.

Submitted by Rebecca Spence

● ● ●

Teen Video Game Contest

Teens have been asking for in-library video game stations since the Mendive Community Library in Sparks, Nevada, started circulating video games. An informal poll of the library's teens was conducted during the planning stages of this program. As a result, a three-hour tournament, including lunch in the park next door, was planned for a Saturday. All of the branch libraries collaborated in purchasing game consoles, and for the actual event the library partnered with the county department of Parks and Open Space and the local Boys and Girls Club.

Playstation 2 game consoles and *Dance Dance Revolution* with dance pads were set up. Participants signed in upon arrival and were set to practicing (in the order of their arrival) at one of the three dance stations. For the first hour, the players cycled through all three stations practicing before keeping score. Hand controllers were available for any teen who could not use the foot controllers.

After lunch break, the actual contest began. Each participant received a scorecard and danced four times, with an official scorekeeper recording their score after each round. One teen volunteer helped keep score. After dancing their four rounds, the teens brought their cards to the score table, and winners were determined based upon highest score. Five library staff, two parks staff, two Boys and Girls Club staff, and one teen volunteer helped with the program.

The game consoles, dance pads, controllers, and games cost around $650, and $60 worth of bottled water was purchased for the event. Each of the library branches contributed money out of their gift funds (from donations and book sales) to buy the hardware and software. Lunch for all the participants was donated by the Boys and Girls Club.

The teens were excited about the program; they did not want it to end. Several teens stayed afterwards to ask when the next game tournament event would be held and provided suggestions for other games that would be fun in a tournament setting. The next dance tournament is planned to be conducted differently. The dancers will be divided into groups according to skill level so there will be winners in each group. Higher skill levels generate higher scores in *Dance Dance Revolution*.

Submitted by Megan Conelly

• • •

Madden 2007 Football Playoffs

Teens in grades seven through twelve attended two evening events at the Ocean County Toms River Branch in Toms River, New Jersey, to play the *Madden NFL 07* game for the Xbox 360. Three plasma TVs were located in the Teen Zone. Microsoft's Xbox 360, wireless controllers, and the *Madden NFL 07* game were needed for this event. A teen was directly involved in the planning. He took registration, promoted the program via a gaming Web site, and helped with setup and cleanup, photo taking, and running the tournament.

Madden NFL 07 is the 17th version of an American football video game. It is available for Xbox, Playstation 2, Nintendo GameCube, Sony PSP, Nintendo DS, Nintendo Game Boy Advance, and the new Xbox 360, Playstation 3, and Nintendo's Wii.

The library partnered with Babbages, a local gaming store in the Ocean County Mall. The Babbages manager provided the Xbox 360 consoles, controllers, games, and prizes. The first-place winner received a new copy of the *Madden 2007* game, second- and third-place winners received limited edition *Madden 2007* faceplates for the Xbox 360. The manager set up the gaming stations and oversaw game settings during the tournament. Registration for the event and promotion was also done at the store.

As teens arrived, they signed in at a table. They were randomly assigned badges that designated the color of the bracket in which they were to play and their number within the bracket. The tournament consisted of three brackets with 16 players in each, modeled after the NCAA men's basketball tournament bracket. They proceeded to the gaming console for their assigned bracket. Players faced off against each other in one-minute quarters, four-minute football games. Winners advanced to the next round until finalists were determined. The semifinal and final games lasted eight minutes, comprised of two-minute quarters.

Gaming guides were available to browse during the tournament. The Babbages manager provided specifications and guidelines for the *Madden 2007* game. Two staff members, three volunteers, and the manager of Babbages managed the tournament.

Quotes from the Teens

"Ya, do it! It brought me into the library. Now I know what a book looks like." – Bellal

"Fun and competitive. You should do it more often." – Joe

"This is fun! These chumps can't handle me." – Shane

Babbages provided all the necessary equipment and materials at no charge.

Future tournaments will allow more time for each game. While it is possible to play one-minute quarters with the *Madden 2007* game, it takes about 15 minutes to actually get through a round. This is due to the players needing to select plays, time-outs, replays, etc. The participants would have enjoyed slightly longer games. They did not seem to mind waiting but enjoyed cheering on friends and others alike.

Submitted by Nicole Politi

● ● ●

Dance Dance Revolution

At the Olathe Public Library in Olathe, Kansas, Teen Advisory Board (TAB) teens and teens attending a hip-hop program suggested a *Dance Dance Revolution* summer program. A teenage staff member and a friend of a staff member coordinated the logistics of "*Dance Dance Revolution* @ Your Library" and collected the equipment. A Playstation console, a video projector, a *Dance Dance Revolution* game, and dance pads were needed.

Dance Dance Revolution is played on a dance pad with up, down, left, and right arrows that are pressed with the feet. The video game screen shows patterns of arrows in rhythm to music and the players must match their moves on the dance pad to the arrows on the screen. The game comes in many different music mixes and is available for GameCube, Xbox, and Playstation 2.

For the first half hour, teens danced with a partner to practice. Practice was important because some of the teens had no experience while others were pros. They were very good about switching off and on, considering the large number of players. After practice time, there was a competition for the #1 dancer on the big pad set. After teens were eliminated, they went and played on the other pad or watched the competition. The teens made decisions throughout the program, which made the program more teen-friendly. There was a lot of laughter and noise. Magazines were available for teens to browse while in the program. One paid staff member, two teen volunteers, and one adult volunteer managed the program.

The only cost for this program was food and drinks. Teen volunteers supplied the DDR equipment. The library had installed an overhead video projector in the meeting room before this program for showing films. After the overwhelming attendance at this program, the Olathe Public Library Foundation presented a grant for purchasing for the library a game console and equipment needed for DDR programs. The library plans to have the DDR program longer and more often and will be doing a DDR and gaming program for Teen Tech Week. Future events will have prizes for the winners and more drinks for the players. If more than one DDR set up is used, the room will be divided better to keep the two consoles separate for noise reasons.

Submitted by Angela Parks

● ● ●

Fantasy Baseball League

Teens aged 11 to 18 at the Peabody Institute Library in Peabody, Massachusetts, form a YA Fantasy Baseball League at the end of March each year, the week before the opening day of the Major League Baseball season. The draft takes place for two hours in the library's YA Drop-In room on the 12 Internet computers, but the draft could be done entirely online. The league lasts for about six months until the end of the regular baseball season in the fall. To manage a team, teens need an e-mail address and access to the Internet throughout the season. After mangers draft their teams, the league is run solely online.

The first time the program ran, none of the library staff was familiar with running a fantasy baseball league. A teen explained the process to the YA librarian. When the teen managers gather for the draft, they help the YA librarian decide on the league's rules. Traditionally, one or more teens also help the YA librarian assemble her team after all the other managers have drafted their players. Once the league begins, the teens are fairly self-sufficient. The YA librarian monitors the league's message boards and responds to trade requests for her team, but she has never had to step in to arbitrate a dispute.

Teens use library computers to create a Yahoo! ID and register their teams. They draw numbers to determine the draft order and then proceed to pick their teams. They record the players that they draft on a sheet that the YA librarian keeps. The librarian also writes down players' names as they are drafted so the teens know who has been "taken." Traditionally, the YA librarian selects her team after the teens have picked theirs. In the days after the draft, the librarian enters the draft results into Yahoo! The rest of the work is done by Yahoo! and the teens. The YA librarian serves as League Commissioner, runs the draft, and manages a team.

The Peabody public schools were involved in advertising the league. They posted flyers and made announcements in the week leading up to the draft. Two to three weeks before the draft, the program was publicized in the library, newspapers, and public schools. The YA librarian created the league online at Yahoo!'s free fantasy baseball service at: http://baseball.fantasysports.yahoo.com/b1. Yahoo! compiles all of the statistics and tracks the teams' standings. ESPN, CBS, and others also host fantasy leagues, many of which are free. The annual *Baseball Prospectus* and several fantasy baseball magazines help teens pick their teams.

The only expenses were purchasing several fantasy baseball magazines for about $15 and a prize at the end of the season for the manager of the winning team, usually a $30 gift card. Usually the prize for this program is provided by the Friends of the Library.

The teens love participating in the annual league. Some teens return every year to manage a team. They enjoy having their own league just for teens. The first year, one teen said at the draft, "I can't believe the library runs a sports thing!" Another teen replied, "We're in a library right now?!? No way!"

Submitted by Melissa Rauseo

● ● ●

Computer Block Party

Teens who do not have the Internet at home enjoy having access at the library and are often reluctant to give up the computers at the end of their allotted times. Invite teens to a Computer Block Party with extended computer time in your library's computer lab or set up a lab in your meeting room with laptops and a wireless Internet connection. Hold the Computer Block Party after hours so teens may socialize without disturbing other patrons.

Chat with the teens who frequent the Internet computers to develop a plan for the program activities. Possible ideas for the event include:

- Internet safety
- Tech share: Share favorite Web sites with a video projector and screen
- *RuneScape*
- Second Life
- YouTube videos
- Survey computer use
- Personal computer time

Draw names for door prizes during free computer time. Tech prizes would be welcome, but so would candy or other snacks if your budget is tight. The event is the optimal time to recruit for a Teen Tech Club and garner ideas for more technology-centered programs.

Submitted by Stephanie Iser

✍ Tech Note ✍

The Entertainment Software Rating Board independently assigns ratings to software games. See www.esrb.org for more information.

- EC is for Early Childhood, age 3 and up
- E is for Everyone, age 6 and up
- E10+ is for Everyone, age 10 and up
- T is for Teens, age 13 and up
- M is for Mature, age 17 and up
- AO is for Adults Only, age 18 and up
- RP is for Rating Pending, used in advertising new releases

✍ Tech Note ✍

Multiplayer Video Games Teens Like:

- *Dance Dance Revolution*
- *Guitar Hero, Guitar Hero 2*
- *Donkey Konga, Donkey Konga 2, Donkey Kong Jungle Beat*
- *Karaoke Revolution*
- *Battlefield 1942*
- *Mario Kart, Mario Kart Double*
- *Super Smash Brothers*
- *RuneScape*
- *John Madden 2007*

● ● ●

Book Display or Bibliography for Gaming Programs

Books and Video Games: Making a Connection

The following is a list of fiction and nonfiction books from The Loft @ ImaginOn that either directly or indirectly address video games or computer games. These books can be displayed to help promote a gaming program, or you may want to refer to this mini-bibliography when trying to find an interesting book for a gamer that frequents your library.

Fiction

Knaak, Richard A. 2004. *The Well of Eternity (WarCraft: War of the Ancients, Book 1)*. New York: Pocket Star.

Niller, Rand, David Wingrove, and Robyn Miller. 2004. *The Myst Reader, Books 1–3: The Book of Atrus; The Book of Ti'ana; The Book of D'ni*. New York: Hyperion.

Seidler, Tor. 2003. *Brainboy and the DeathMaster*. New York: Laura Geringer.

Weiss, D. B. 2003. *Lucky Wander Boy*. New York: Plume Books.

Wieler, Diana. 1995. *Ranvan: A Worthy Opponent*. Toronto: Groundwood Books.

Graphic Novels

Andreyko, Marc, and E. J. Su. 2005. *Castlevania: The Belmont Legacy*. San Diego, CA: IDW Publishing.

Gallagher, Michael et al. 2003. *Sonic the Hedgehog: The Beginning*. Mamaroneck, NY: Archie Comics.

Hamazaki, Tatsuya, and Rei Izumi. 2003. *.hack//Legend of the Twilight*. Los Angeles, CA: Tokyopop.

Holkins, Jerry, and Mike Krahulik. 2006. *Penny Arcade, Volume 1: Attack of the Bacon Robots!* Milwaukie, OR: Dark Horse Comics.

Jurgens, Dan et al. 2002. *Tomb Raider, Volume 1: The Saga of the Medusa Mask*. Los Angeles, CA: Image Comics.

Knaak, Richard A., and Kim Jae-Hwan. 2005. *Warcraft: The Sunwell Trilogy*. Los Angeles, CA: Tokyopop.

Murakami, Maki. 2004. *Gamerz Heaven*. Houston, TX: ADV Manga.

Oprisko, Kris, and Ashley Wood, 2005. *Metal Gear Solid, Volume 1*. San Diego, CA: IDW Publishing.

Pak, Greg et al. 2005. *Marvel Nemesis: The Imperfects Digest*. New York: Marvel Comics.

Waid, Mark et al. 2005. *City of Heroes, Volume 1*. Los Angeles, CA: Image Comics.

Poetry

Barkan, Seth. "Fingers' Flynn." 2004. *Blue Wizard Is About to Die: Prose, Poems, and Emoto-Versatronic Expressionist Pieces about Video Games 1980–2003*. Las Vegas, NV: Rusty Immelman Press.

Non-Fiction

Brady Games. 2006. *Secret Codes. 2006, Vol. 1: PlayStation 2, Xbox, Gamecube, Nintendo DS, PSP, Gameboy Advance*. Indianapolis, IN: BradyGames.

DeMaria, Rusel, and Johnny L. Wilson. 2004. *High Score!: The Illustrated History of Electronic Games, 2nd Ed.* New York: McGraw-Hill Osborne Media.

Ferriere, Gala, and Jim Zubkavich. 2005. *Street Fighter: Eternal Challenge*. Chicago, IL: Devil's Due Pub.

Kent, Stephen L. 2001. *The Ultimate History of Video Games: From Pong to Pokemon: The Story behind the Craze That Touched Our Lives and Changed the World*. Roseville, CA: Prima Pub.

Ozawa, Tadashi. 2000. *How to Draw Anime & Game Characters, Volume 1–5*. Tokyo: Graphic-Sha.

Pardew, Les. 2004. *Game Art for Teens (Game Development Series)*. Boston, MA: Thomson Course Technology/Premier Press.

Prima Games. 2005. *Codes & Cheats, Fall 2005 Ed*. Roseville, CA: Prima Games.

Prima Games. 2006. *Codes & Cheats, Winter 2006 Ed*. Roseville, CA: Prima Games.

Prima Games. 2003. *The Ultimate Code Book: Cheats and the Cheating Cheaters Who Use Them*. Roseville, CA: Prima Games.

Prima Games. 2005. *Video Game Cheat Codes*. Roseville, CA: Prima Games.

Sethi, Maneesh. 2003. *Game Programming for Teens*. Boston, MA: Thomson Course Technology/Premier Press.

Further Reading on Gaming and Libraries

Library Success Wiki—Gaming Page. [Online] Available at: www. libsuccess. org/index.php?title=Gaming

Beck, John C., and Mitchell Wade. 2004. *Got Game: How the Gamer Generation Is Reshaping Business Forever*. Boston, MA: Harvard Business School Press.

Gee, James Paul. 2004. *What Video Games Have to Teach Us about Learning and Literacy*. New York: Palgrave Macmillan.

Johnson, Steven. 2005. *Everything Bad Is Good for You: How Today's Pop Culture Is Making Us Smarter*. New York: Riverhead.

Prensky, Marc. 2006. *Don't Bother Me Mom I'm Learning: How Computer and Video Games Are Preparing Kids for Learning*. St. Paul, MN: Paragon House.

Chapter 2

The Internet Presence Connection

Utilizing multimedia content, libraries can establish an exciting Internet presence. Involving teens in the planning stages yields awesome results. Podcasting, MySpace, Flickr, Blogs, Wikis, RSS feeds, polls, and other creative and interactive site elements will attract teens. Promote programs and library materials and communicate with teens through the library's Web site and social networking sites like Second Life. Invite interested teens to help design the teen area of the library's Web site.

Teen Podcast

The Cheshire Public Library in Cheshire, Connecticut, produces a podcast that is a teen-driven cultural magazine with listeners of all ages all over the world. Each episode is a variety program anchored by a teen host. Program segments may include poems, book reviews, music reviews, comedy sketches, or songs; the format is flexible. A teen editorial board works to recruit submissions and select segments and follows each episode through to release. These teens, in grades nine through twelve, supervise the show, but also often contribute their talents "on air" as well as behind the scenes.

✍ What Is a Podcast? ✍

A podcast is a multimedia file available on the Internet. Podcasts are often audio files that can be downloaded to portable devices like MP3 players or a desktop PC. Audacity is a free recording software program you can use to create your own podcasts. Audacity is available at: http://audacity.sourceforge.net/ Gabcast

25

is another site where you can create podcasts and audio files for blogs, at: www.gabcast.com. Garage Band software is used for MAC computers.

Garage Band software (www.apple.com/ilife/garageband/) and a plug-and-play microphone connected to a Macintosh computer are used to create the podcasts. The podcast is hosted by the Internet archive and made available on the library's Web site at: www.cheshirelibrary.org/teens/cplpodcast.htm and at the iTunes music store. To reach a wider audience, the CPL Podcast is available several different ways. Directions on how to subscribe to the podcast through the iTunes music store and links to the streaming audio for each episode are on the Web site. Each episode is also burned on CD and made available at the library. Each CD includes directions on how to find the podcast online and an information sheet on how to get involved with the program. Statistics, including the number of downloads, are recorded by Feedburner, a free service, at www.feedburner.com. To date, Feedburner shows over 1,300 downloads.

The podcast is funded through the generosity of the Friends of the Cheshire Library. Friends' funds purchased a microphone for $150 and also go towards promotional materials. The podcast is a low-cost high-tech program.

Teens are recruited as participants for the podcast from high school clubs and classes, including the graphic design class, the drama club, and Writer's Block, a writing club. Teens give great feedback about the program to the local press. The *New Haven Register* ran a quote by a teen member of the editorial board: "I've been waiting for something like this my whole life. It's something new, something interesting, something that gets a lot of people involved."

Teens are drawn to the project because they love the satisfaction they feel as each episode is released. The same teens continue participating because they feel a sense of ownership and creative control. They constantly tweak the format, the procedure, and the content as they strive to produce a better podcast.

Submitted by Sarah Kline Morgan

Figure 2.1: Cheshire Public Library Podcast Logo

Cheshire Public Library

Designed by one of the Cheshire Public Library teens, the podcast logo adorns the teen page of the library's Web site available at: www.cheshirelibrary.org/ teens/cplpodcast.htm. All the podcast episodes are available for listening at this site. (Reprinted with permission of the Cheshire Public Library.)

✍ Sites for Podcasting ✍

Check out these two sites for more information about podcasting:

Wikipedia. Available at: http://en.wikipedia.org/wiki/ Podcasting

Podcasting FAQs. Available at: www.podcastingnews.com/ topics/Podcasting_FAQ.html

---

Libraries with Podcasts

Looking for podcasting ideas? Check out a few podcasts from other libraries:

Hennepin County Library offers teen movie and book reviews at: www.hclib.org/teens/Podcasts.cfm

(cont'd.)

Libraries with Podcasts *(Continued)*

The New York Public Library hosts topical discussion podcasts at:
http://teenlink.nypl.org/turnitup.html

The Public Library of Charlotte & Mecklenburg County offers podcasts of news, programs, and commentary at:
www.libraryloft.org/podcasts.asp

East Oakland Community High School podcasts discuss teen topics at:
www.eastoakland.libsyn.com/

The Library Success Wiki lists many more libraries offering podcasts at:
www.libsuccess.org/index.php?title=Podcasting

● ● ●

Library MySpace Page

The Haverhill Public Library in Haverhill, Massachusetts, created a library's MySpace presence at the request and with the help of local teens in April 2006. Once the director approved the idea,

Figure 2.2: Haverhill Public Library MySpace Screenshot

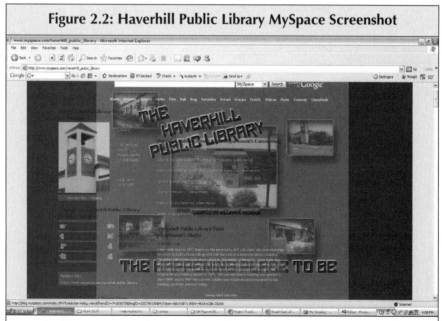

Haverhill Public Library connects with their teens on MySpace with a page designed with teen input. (Reprinted with permission of the Haverhill Public Library.)

the librarian approached a couple of the teen regulars and asked them to assist in creating the library's initial page at: www. myspace.com/haverhill_public_library

A teen that attended the local technical high school in the graphics program created a background, which he designed and coded himself, with input from the librarian, to make the space cool but appropriate for the library. The page is currently the second background created entirely by another local teen who also took photos to add to the page.

The MySpace page is used to inform teens about upcoming programs and library closings. The MySpace blog highlights new books, book reviews, book award news, read alike lists, etc., and the librarian tries to blog something every day so the teens get into the habit of checking in at the library's page on a regular basis. The current reading feature is used, as well. The MySpace page is linked to the library's main site, the teen Web page, and the library catalog.

Teens think that it is really cool that the library has a MySpace page and are very excited when they end up in the top of the friends list. Volunteers and library regulars hold permanent places at the top of the friends list, while local teens rotate through the rest of the top half of the Library's friend list. The bottom half of the list consists of popular teen authors who have MySpace pages.

Submitted by Alissa Lauzon

Libraries with MySpace

Looking for ideas for your MySpace page? Check out the pages these libraries have created:

Coshocton Public Library Animanga Club at:
www.myspace.com/cplmangaclub

Teen Librarians @ Glendale Main at: www.myspace.com/teenlibrariansgpl

YA Space at Madison Public Library at: www.myspace.com/yaspace

Cool Librarians at: www.myspace.com/sopl

The Library Success Wiki lists many more libraries offering MySpace pages for their teens at: www.libsuccess.org/index.php?title=MySpace_%26_Teens

✍ What Is a Web Feed? ✍

A Web feed delivers frequently updated content to subscribed users in a program similar to e-mail. Distributors syndicate a Web feed and users subscribe to it. Subscribers use an aggregator program, or feed reader, to check sites on the subscription list for new content periodically. Atom feeds and RSS feeds are two formats of Web feeds. The acronym "RSS" has a few different translations: Really Simple Syndication, Rich Site Summary, Resource Description Framework Site Summary. Offering an RSS feed on your Web site or blog will help teens stay up to date with new content without having to remember to visit your site.

✍ What Is Flickr? ✍

Flickr is a photo-sharing site. Users can store photos at the site, share them in albums, and link the album to blogs and Web sites. Photos can be organized with tags and insets, and viewers can subscribe with RSS or Atom feed readers.

● ● ●

MediaSpace Event

Teens enjoy participating in social networking sites on the Internet, but many may not have the means to add exciting multimedia elements to their pages. A MediaSpace event at your library will give them the opportunity to create podcasts, videos, and digital photos.

The Teen Advisory Board can help create the rules and logistics of the program. Length of videos and podcasts and number of photos and rules for content need to be established. Set up a booth for each media type and staff it with volunteers to help facilitate creating files. A digital camera with a suitable backdrop is needed for the digital photo booth. The podcasting station needs a computer, microphone, and podcasting software, such as Audacity at: http://audacity.sourceforge.net. A digital video camera and a backdrop are needed for the video booth. Provide alternate activities, such as video and board games, for teens waiting their turns. Preregistration and appointments would help the process along if enough volunteers are not available.

Collect contact information at the program along with some descriptive information and type and number of files created for each teen. After the program, the files can be compressed and burned onto CDs and given to the teens to upload to their Facebook or MySpace pages. Smaller files can be e-mailed to the teens.

A MediaSpace event can be a fun follow-up to an Internet Safety presentation for teens and parents. Teens who have not yet created Web sites would benefit from a handout with information about popular sites like MySpace, YouTube, OurMedia, or Flickr. Have permission slips available if any of the files will be used on the library's Web site.

Submitted by Stephanie Iser

Further Reading on Creating an Internet Presence

Apple. "Podcasting in Education." [Online] Available at: www.apple.com/education/solutions/podcasting/

Colombo, George, and Curtis Franklin. 2005. *Absolute Beginner's Guide to Podcasting*. Indianapolis, IN: Que.

Eash, Esther Kreider. "Podcasting 101 for K–12 Librarians" [Online] Available at: www.infotoday.com/cilmag/apr06/Eash.shtml

Morris, Tee, and Evo Terra. 2005. *Podcasting for Dummies*. Hoboken, NJ: Wiley.

Wikipedia. "MySpace" [Online] Available at: http://en.wikipedia.org/wiki/Myspace

Chapter 3

The Audiobook and Music Connection

"Wired and connected" is a literal description of the teens grooving along with iPods clipped to their clothes and earbuds in their ears. What are they listening to? Music, books, even podcasts—any audio file can be loaded into these devices. Contribute to the audio flow of entertainment and information going in by providing downloadable files. Cater to your teens' listening needs by creating a space on the library Web site where teens can post their Top Ten favorite songs of the moment. Teens can also create playlists that connect to their favorite books, manga, graphic novels, movies, and games. Post the lists of songs on the library Web site. Introduce your teens to audiobooks in all formats by playing one during a craft program.

MP3 Circulation

The Lake Forest Academy Library in Lake Forest, Illinois, is working to meet the challenge of matching library circulation techniques with modern technology formats. Audio recordings can be easily compressed into digital files, and CD collections may be replaced by MP3 collections to meet the popularity of audio materials without the burden of maintaining a traditional physical collection. The challenge remains to establish a circulation and cataloging model to incorporate digital audio formats into the collection and to circulate the material by transferring material from library computers to patrons' iPods. The Lake Forest Academy (LFA) has begun the process by focusing on audiobooks and recorded music.

Literature teachers frequently recommend audio recordings of Shakespearean dramas for ESL students and other students who may find the language too difficult to read. The LFA music department often asks to have particular recordings placed on reserve. Students are required to listen to reserved works on CD in the library. However, neither audiobook CDs nor music CDs circulate very well at LFA since MP3 players are quickly replacing portable CD players for audio playback.

An iPod Nano was purchased in the spring of 2006 by the LFA and began circulating. Circulating a Nano begs the question of how to circulate both hardware and content, as students would essentially be "checking out" a title or titles to the player itself. A copy of the BBC Radio 3's production of Shakespeare's *Romeo and Juliet* was purchased at the iTunes music store and transferred to a laptop's iTunes application folder. This title was then uploaded to the Nano and made available for in-house checkout to students at the beginning of the fall semester in 2006. The Nano did not circulate well, which suggested two things:

1. Students were not interested in checking out a device that could not leave the library no matter how familiar or accessible.
2. Students were not interested in checking the device out for the sole purpose of listening to *Romeo and Juliet*.

Each suggestion then prompted a response:

1. Students might be interested in checking out digital content to their own players.
2. Students might explore the library's music collection more if it were in MP3 format.

✍ Tech Note ✍

Zoomerang is one of the online survey software sites that can be added to an existing Web site or e-mailed to patrons. Zoomerang provides templates, or you can create your own, and it compiles the results in a report. A free basic survey is available that includes up to 30 questions and up to 100 responses per day. Learn more about Zoomerang at: http://info.zoomerang.com/

Figure 3.1: Zoomerang Survey Results

1. Do you own an iPod?			
Yes		108	77%
No		32	23%
Total		140	100%

2. Do you own another type of MP3 player (other than an iPod)?			
Yes		32	23%
No		108	77%
Total		140	100%

3. Would you use an iPod/MP3 player to listen to an assigned book?			
Yes		98	71%
No		41	29%
Total		139	100%

4. Would you use your iPod/MP3 player to check out library audio materials (including music)?			
Yes		114	81%
No		26	19%
Total		140	100%

5. Are you male or female?			
Male		71	51%
Female		69	49%
Total		140	100%

6. Class?			
Freshman		38	27%
Sophomore		34	24%
Junior		30	22%
Senior		38	27%
Total		140	100%

Zoomerang collected and calculated the results of a survey of teens about their use of MP3 players. (Reprinted with permission of the Lake Forest Academy Library.)

The Zoomerang survey confirmed that most of the students owned iPods, that they would be interested in checking out both music and audiobooks to iPods, and that a user-centered approach to MP3 circulation was preferable to the limited model associated with circulating multiple library-owned iPods.

Technical issues had to be solved before launching the iPod/iTunes circulation program. iPods formatted on Macintosh computers will not communicate with an iTunes application on a Windows PC. For students owning Macs, communication had to be established between Mac-formatted iPods and the library's Thinkpad laptop. A product from www.Mediafour.com known as Xplay2.2 was purchased for the circulation desktop and the librarian's laptop. Xplay 2.2 is a third-party client designed to allow cross-platform communications between iPods and iTunes applications.

The practice of buying and circulating MP3s in libraries is relatively new, so full MARC records in a standard format were not available from OCLC. LFA's cataloguer used available audio/visual records as a template for creating an LFA library standard record for MP3 files.

The iTunes application defaults to automatic iPod updating when the device is connected to the computer, a setting that replaces what was previously on the device with what is currently in the user's iTunes library. LFA felt the loss of user-owned content (no matter how temporary) was unacceptable. By changing application preferences to manual updating in both iTunes and Xplay2.2, it was possible to circumvent the problem of content loss by adding and removing library-created playlists to each iPod. Students are more likely to explore what the library has to offer by way of their iPods if they do not have to sacrifice already loaded playlists. The decision was made to circulate both types of playlists for a period of one week consistent with the library's CD circulation policy.

The librarian compiled a list of 50 recordings, both academic and popular, that best reflect the intellectual crisis and eventual split in twentieth-century music. This list, to which titles are routinely added, serves as a guide for what will become the new music collection. Faculty also recommended audiobook titles to address the needs of the curriculum.

A PowerPoint presentation complete with synchronized audio played during an all-school meeting to launch the new resources. This approach was an attempt to reach the students more directly and in a manner suitable to the introduction of such a technology-oriented program. The response was tremendous, and the buzz around campus continued throughout the following week.

Teachers and students alike check out the music and audiobook playlists. In addition to the excitement and convenience of having materials added to their own iPods without having to disrupt their own playlists, LFA patrons have also begun to check out the four library-owned iPods.

Submitted by Grier Carson

What Is an MP3?

MP3, short for MPEG-1 Audio Layer 3, refers to a compressed audio file. Using an MP3 standard greatly reduces the size of the audio file with little or no loss of sound quality. MP3 files can be downloaded onto personal MP3 players.

An iPod is one brand of MP3 players available. iPods can also play movies, TV shows, and games as well as audio files. iPod Nano, mentioned in the previous program description, is a slimline iPod just for audio files. iPod also offers a clip-on model called iPod Shuffle. It is the least-expensive iPod, holding up to 240 songs. iPods are great ideas for teen program grand prizes.

● ● ●

Playaway Audiobooks

The latest audiobook phenomenon to hit the land of public libraries is the Playaway. The device (2⅛ in. by 3⅜ in. and ¼ in. thick) resembles an iPod and comes with one book title per unit. It runs on one AAA battery and is attached to a lanyard for convenient hands-free listening. At the Holmes County Public Library in Millersburg, Ohio, patrons may use their own earphones or purchase earbuds from the library for $1. The library's policy is to provide one new battery for each checkout.

The library purchased about three dozen titles to test the circulation. Six circulated the first month and 48 during the second month. Now they are never on the shelf! The teens have become enamored with the devices.

Teen Quotes

"There is no extra case or container to take along when you need to change CDs. It's all right there. It's cool!" commented one teenage user.

"They are really easy to use and you can listen to them while you're doing something else," says Christen, a new Playaway fan.

You don't have to be a technology whiz to operate the Playaway, just push the play button and voila! You're being read to! The appeal is in the size (it fits in your pocket) and the fact that there is no inserting or changing of CDs. Even better, there are no CDs to be scratched and, therefore, no replacement CDs to order. Prices begin at $29, so they are very comparable to audiobook titles on CD. The patron can also purchase an FM transmitter from the Playaway Web site or use another FM transmitter to listen through the car radio.

The only negative comments have related to managing the control buttons. It takes two pushes to start the player: One to turn the unit on and another firm push to start the recording. Once the patron is familiar with the operation, it is easy to use.

More and more teens at the library are listening to audiobooks, so the choice to try this format was not difficult. Titles like *Eragon* and *The Da Vinci Code*, which circulate well in print format, were purchased in an attempt to attract teen patrons right off the bat. For more information on the Playaway devices, go to: http://store.playawaydigital.com/

Submitted by Linda Uhler

✍ Tech Note ✍

Your library may decide to sell earbuds for Playaways as Holmes County does. In Ohio, OhioNet sells them for about $.71 each at: www.ohionet.org/libprods_new.shtml. BWI sells 50 in a bag for $39.50 at: www.bwibooks.com/

Further Reading on Audio Products

Biersdorfer, J. D. 2006. *iPod & iTunes: The Missing Manual, Fourth Ed.* Sebastopol, CA: O'Reilly Media.

Honan, Mathew. 2006. "Libraries Turning to iPods and iTunes." [Online] Available at: http://playlistmag.com/features/2006/02/library/index. php

Stephens, Michael. 2005. "The iPod Experiments." *Library Journal.* April 15 [Online] Available at: www.libraryjournal.com/article/CA515808.html

Chapter 4

The Art and Film Connection

Creating art and movies is a fun and creative pastime for teens. Technology brings this creativity to a new level. Computer software can be used to create and enhance artwork and photography, while small affordable video cameras have made filmmaking accessible to everyone. Teens love recording their friends' antics and enjoy producing a real film for an audience, sometimes with a script. YALSA offered a Teen Tech Week Video Contest, and teens had uploaded their films to YouTube for judging. Consider a People's Choice contest in your library to judge teens' film or digital photo creations. Libraries are creating promotional films with their teens for their Internet sites. Watching commercial movies and original films also makes a fun teen program. A movie license can turn your meeting room into a movie theatre. Just add popcorn!

Night at the Movies

Holmes County Public Library in Millersburg, Ohio, shows recent movies in their meeting room during the "Teen Night at the Movies" program, attracting many teens who would otherwise need to leave the county to go to a theatre. Teens, ages 12 to 18, are invited to the free monthly showings. The library's Teen Advisory Board chooses the movies to show. For the December movie, the library charges one nonperishable food item per person, which is then given to the local Share-A-Christmas charity.

The library acquired a site license to show movies. The equipment setup to show films on DVD includes a DVD player or laptop computer with a DVD player, a large movie screen or a plain light colored wall, a video projector, and a sound system. The teen programming budget covers the $20 expense per movie night for soda, popcorn, and popcorn bags.

Fifteen minutes prior to the start of a movie, a PowerPoint program advertises the library's upcoming teen programs and services, in imitation of the previews seen at a movie theatre. Books and other materials are on display with a sign that says, "If you like this movie, you might like this book." Teens love being able to see free movies with their friends.

Submitted by Linda Uhler

✍ Movie Licensing ✍

A movie license is required to show films in the library. Two popular companies are Movie Licensing USA at www.movlic.com/library/library.html and Motion Picture Licensing Corporation at www.mplc.com. License fees for libraries are usually based on the number of active patrons.

Movie Licensing USA is a licensing agent for Walt Disney Pictures, Touchstone Pictures, Hollywood Pictures, Warner Bros., Columbia Pictures, TriStar Pictures, Paramount Pictures, DreamWorks Pictures, Metro-Goldwyn-Mayer, Universal Pictures, Sony Pictures, United Artists, and various independent studios, and provides the Movie Copyright Compliance Site License. Send an e-mail to mail@movlic.com to inquire about cost. There are separate telephone numbers for schools and public libraries. For Movie Licensing USA for schools, call (877) 321-1300 (toll-free). For Movie Licensing USA for public libraries, call (888) 267-2658 (toll-free).

Motion Picture Licensing represents over 60 producers and distributors, including Walt Disney Pictures, Warner Bros., Scholastic Entertainment, McGraw-Hill, Sony Pictures Classics, Tommy Nelson, and World Almanac, and provides an Umbrella License[SM]. Call (800) 462-8855 or send an e-mail to info@mplc.com.

● ● ●

Video Production Workshop

Springfield-Greene County Library in Springfield, Missouri, invited teens with an interest in filmmaking to a series of programs to learn how to make their own videos. The series was held on four Saturday mornings in January and February of 2006 in time to produce a film and enter it in the local film festival. Each session lasted two hours, from 10 am to noon, and was held in a large meeting room at the system's primary branch.

The local film group, Missouri Film Alliance of Springfield, put together the workshop, provided the expertise and specialty equipment, and ran the program. Their annual film festival includes a teen category, and the group wanted to encourage participation and awareness by providing instruction in a relaxed atmosphere.

This was a basic video production workshop to help teens become better filmmakers and storytellers. The sessions used a computer and projector, various handheld video cameras, film editing computer software, an editing machine, and a sound mixing board. Each session built upon the previous ones. The teens were introduced to different elements of filmmaking, given a chance to work on their skills during the session, then encouraged to apply the knowledge to their own films during the week between sessions.

In the first session, Story Development, participants learned the basics of script format, character and plot development, and strategies for dialogue. In the second session, Pre-production Concerns and Activities of the Director, the teens prepared a script for production and learned techniques for casting, selecting locations, and creating storyboards. In the third session, Pre-production Concerns and Basic Lighting and Sound Techniques, the group explored the technical aspects of the film with lighting, camera, and audio production work. In the final session, Post-production Concerns and Editing as Storytelling, the participants were introduced to principles of continuity and montage editing and learned different editing strategies such as cutting dialogue and action scenes. A week or two later, one of the completed teen films was screened prior to the regular after-hours teen lock-in.

A couple of library staff members made arrangements for publicity, room setup, and computer use. The sponsoring group provided several volunteers for each session. A half-dozen key volunteers came regularly and planned the program. A small portion of staff time, 60 to 90 minutes maximum, was required. The equipment was either on hand or provided by the Film Alliance.

The group also provided a number of photocopied handouts. The total cost was less than $75.

Generally, the comments were positive, although there were several complaints from teens who were not able to attend. They requested a repeat of the program. One of the Teen Library Council members was particularly excited about it and showed her work right before the monthly teen lock-in.

The workshops will be repeated in early 2007 and have been expanded to six sessions over the course of eight weeks, with more emphasis on story development and basic sound and lighting techniques. The age range has been made 15 to 18 for a variety of reasons, including the likelihood of promoting the program to the media departments of local high schools. All completed projects will be screened at the Third Annual Show Me Missouri Film Festival to be held in the spring of 2007. Registration is required and has been limited to 20 at the request of the presenters. A resource list will be posted on the teen Web page to be printed. The teen blog will have entries about the program, the film festival, and filmmaking resources.

Submitted by Beth Snow

● ● ●

A Week of Japanese Entertainment

Teens were invited to a week of Japanese entertainment during spring break at Elizabeth Public Library in Elizabeth, New Jersey. A two-hour program was held each day in the library's auditorium, planned by the Teen Advisory Council.

Each evening during the week a full-length anime feature was shown. A DVD player and LCD projector were used, and the DVDs, selected by the teens, were purchased by the library to be added to the circulating collection. On Friday, a Manga Swap and Discussion was held using the library's manga collection and books brought in by the teens. A karaoke program was held on Saturday. The music was chosen by the teens, and a local Korean deli prepared California rolls to serve during the karaoke program.

The expenses were minimal; about $50 was spent on food and approximately $80 on DVDs. The food money came out of the programs budget and the DVD money came from the AV collections budget. The karaoke machine and music were borrowed.

The teens loved the program. The week was so successful, an Anime Club was started, which now boasts upwards of 25 members at any given meeting. The library plans to repeat the program, adding a Basic Japanese class, a chopsticks tutorial, and cosplay—anime- and manga-themed costume parties.

Submitted by Kimberly Paone

● ● ●

Digital Graphic Art Contest
The Teen Techies group at Homer Township Public Library in Homer Glen, Illinois, suggested inviting high school students to enter a digital graphic arts contest during the month before Teen Tech Week. The group also named the program, and one member designed the logo. Graphics software such as Photoshop or Fireworks were used, and a $25 gift certificate was the grand prize. Entries for a digital art contest were requested as book covers, CD or DVD covers, or movie or concert posters.

Figure 4.1: Digital Imagination Artist Release Form

Artist Release Form

Statement 1

The owners/copyright holders of this digital graphic artwork grants permission to the Homer Township Public Library District to use the artwork in the following manner:

The Homer Township Public Library District may reproduce the digital graphic artwork, in whole or in part, in their displays, publications, web pages, and presentations for an unlimited period of time.

The Homer Township Public Library District will include artist credit in their identification material on the digital graphic art and will endeavor to see that proper credit is given to the artist by all media.

I hereby waive my right to inspect and/or approve the finished product(s).

I affirm that the digital graphic art submitted is original on my part and that I am sole owner of the material, and that neither the material nor the permission granted hereby infringes upon the rights of others.

(cont'd.)

Figure 4.1: Digital Imagination Artist Release Form *(Continued)*

Statement 1 *(cont'd.)*
I understand the artwork will not be returned.

Signed: _____

Print name: _____

Date: _____

Statement 2: Participation of Minors
The signature of a parent or legal guardian is required for participation. I hereby give my permission for the participation of the named minor in the digital graphic art contest. I have read and agreed to Statement 1.

Name of Minor: _____

Signature of Minor: _____

Parent/Guardian's Name: _____

Parent/Guardian's Signature: _____

Relationship to Minor: _____

The rules for the Digital Imagination Graphic Art Contest. (Reprinted with permission of the Homer Township Public Library.)

Figure 4.2: Digital Imagination Art Contest Rules

A Digital Imagination

Digital Graphic Art Contest Rules

- Digital graphic art should follow the "A Digital Imagination" theme.
- Participants are limited to teens in 9th–12th grade.
- Submit one digital graphic art piece only.
- Digital graphic artwork is art created on a computer in digital format.
- Digital artwork may not contain copyrighted work.
- Digital graphic artwork must be high resolution and submitted in .jpg, .gif, or .png format.
- Submissions may be a hard copy or a digital copy.
- Entries must be submitted between February 1, 2007 and March 9, 2007.
- Artists must submit their entries to Homer Township Public Library

(cont'd.)

Figure 4.2: Digital Imagination Art Contest Rules *(Continued)*

Digital Graphic Art Contest Rules *(cont'd.)*

- The library will have one winner. The winning artist will receive a $25 gift certificate to Borders.
- Digital graphic artwork may be placed on display in the library.
- Names of the digital graphic artist (and any subjects) may be used in promotional materials.
- The digital graphic artist must not break any copyright laws. It is the digital graphic artist's responsibility to determine if any laws will be broken.
- Obscene digital graphic artwork will not be tolerated. The library will take action if an entry is determined inappropriate.
- Contest is limited to Homer Township Public Library District patrons.
- Entries that do not comply with these rules will be disqualified.

Hard Copy (disks, CD-ROMs):
- Attach a copy of the entry form to hard copy.
- Submit a signed copy of the digital graphic artist release form.
- Submit a signed copy of the digital graphic subject release form from EVERY person shown in the artwork.
- Forms and artwork must be submitted all at the same time.
- Incomplete entries will not be accepted. It is the teen's responsibility to make sure entries are complete.

Email submissions
- Email the photograph as an attachment to Homer Township Public Library District (homerlibrary@gmail.com).
- Print out a copy of the sent email.
- Attach a copy of the entry form to the email printout as verification.
- Submit a signed copy of the digital graphic artist release form.
- Submit a signed copy of the digital graphic subject release form from EVERY person shown in the artwork.
- Forms and artwork must be submitted all at the same time.
- Incomplete entries will not be accepted. It is the teen's responsibility to make sure entries are complete.

Homer Township Public Library District
14230 W. 151st St. ● Homer Glenn, IL 60491 Phone: (708) 301-7908

The Digital Imagination Graphic Art Contest Art Release form. (Reprinted with permission of the Homer Township Public Library.)

Figure 4.3: Digital Imagination Subject Release Form

Digital Graphic Subject Release Form

All individuals pictured in submitted digital graphic artwork must sign a copy of the following consent statement.

Statement 1

I hereby consent that the digital graphic artwork, in whole or in part, containing my image for the Digital Imagination contest may be used by the Homer Township Public Library District for displays, publications, web pages, and presentations for an unlimited period of time.

The Homer Township Public Library District will include artist and subject(s) credit in their identification material on the artwork and will endeavor to see that proper credit is given to the artist and subjects by all media.

I hereby waive my right to inspect and/or approve the finished product(s).

Signed:_____

Print name:_____

Date: _____

Statement 2: Participation of Minors

The signature of a parent or legal guardian is required for participation. I hereby give my permission that the artwork, in whole or in part, of the named minor may be used in the digital graphic artwork contest. I have read and agreed to Statement 1.

Name of Minor: _____

Signature of Minor:_____

Parent/Guardian's Name:_____

Parent/Guardian's Signature: _____

Relationship to Minor: _____

The Digital Imagination Graphic Art Contest Subject Release form. (Reprinted with permission of the Homer Township Public Library.)

Figure 4.4: Digital Imagination Entry Form

A Digital Imagination

Have your winning digital graphic artwork featured in the library and on the library's teen webpage! Digital graphic artwork is art created on a computer in digital format.

The winning artist will receive a $25 gift certificate to Borders.

Entry deadline is 3/9/07.

Release forms are available at the library and online.

How to participate:

- Submit one digital graphic art piece only. Submissions must be a digital copy. All copies must be high resolution and in .jpg, .gif, or .png format.
- Submit a copy of the entry form.
- Submit a signed copy of the artist release form.
- If needed, submit a signed copy of photo subject release form from ALL subjects shown in photograph.
- Drop off entry at your local library between February 1 and March 9, 2007.

Incomplete entries will not be considered.

Complete details available at www.homerlibrary.org/teenevents.asp

Name _____

Title _____

Grade _____

Phone _____

Address _____

The Digital Imagination Graphic Art Contest entry form. (Reprinted with permission of the Homer Township Public Library.)

Submitted by Alexandra Tyle

Further Reading on Filmmaking

Fitzsimmons, April. 1997. *Breaking & Entering: Land Your First Job in Film Production.* Los Angeles: Lone Eagle.

Harmon, Renee, and Jim Lawrence. 1997. *The Beginning Filmmaker's Guide to a Successful First Film.* New York: Walker and Co.

Knox, Dave. 2005. *Strike the Baby and Kill the Blonde: An Insider's Guide to Film Slang.* New York: Three Rivers Press.

Lanier, Troy, and Clay Nichols. 2005. *Filmmaking for Teens: Pulling off Your Shorts.* Studio City, CA: Michael Wiese Productions.

LeKich, John. 2002. *Reel Adventures: The Savvy Teens' Guide to Great Movies.* Toronto: Annick Press, distributed by Firefly Books.

Nanda, Jai. 2003. *I Know What You Quoted Last Summer: Quotes and Trivia from the Most Memorable Contemporary Movies.* New York: St. Martin's Griffin.

Newton, Dale, and John Gaspard. 2000. *Digital Filmmaking 101.* Studio City, CA: Michael Wiese Productions.

Parkinson, David. 1995. *The Young Oxford Book of the Movies.* New York: Oxford University Press.

Roeper, Richard. 2003. *Ten Sure Signs a Character Is Doomed and Other Movie Lists.* New York: Hyperion.

Shaner, Pete, and Gerald Everett Jones. 2004. *Digital Filmmaking for Teens.* Boston, MA: Thomson Course Technology Professional Trade Reference.

Chapter 5

The Reading Program Connection

Combining reading and technology can be a useful tool to connect teens with books. With prizes, online registration, activity programs, competitions, and entertainment, technology can add the "pizzazz" that will attract teens to your reading program. Libraries can extend their summer reading programs (SRPs) to teens stuck at home through online participation. Try an online book review message board, online polls, authors chats, inserting title links to the catalog for quick reserves, and sharing your own young adult reading choices. Promote your in-library summer reading programs with teen-made podcasts and videos on the library Web site.

Music-Themed Summer Reading Program

Unplugged @ Your Library—READ was a 10-week music-themed summer reading program for teens at Cuyahoga County Public Library's 26 branches near Cleveland, Ohio. From prizes to programs, technology helped make music one of their most exciting and popular summer reading themes.

The program was launched in the previous fall with a contest to design a logo to be used throughout the summer. The winning art was created by a local high school student. Each branch received a large cardboard guitar to promote the program and was encouraged to be creative and involve the teens in decorating the guitar.

A committee coordinated high school bands from the Live Nation High School Rock-Off in January to perform live concerts at various branches for the Unplugged @ Your Library Rock-Off

Band Tour Summer 2006. Live Nation offered discounts on the High School Rock-Off CD that was used for prizes and incentives. They supplied band contact information for the library concert tour. The International House of Blues Foundation presented the Blues School House outdoor concert tour at five branches.

Figure 5.1 Unplugged Logo

The teen-designed logo helped launch the Unplugged @ Your Library music-themed summer reading program. (Reprinted with permission of the Cuyahoga County Public Library.)

The first 50 teens who registered for the Unplugged @ Your Library—READ SRP received a special High School Rock-Off CD with SRP instructions and a chance to be an instant winner. Various songs from the rock-off were also on the CD, making it a keeper. Weekly prize drawings were held. Creative MP3 players, a behind-the-scenes tour at the local radio station KISS-FM with a local DJ, and tickets to a Buzzard Lounge concert from WMMS were grand prizes. Chipotle, a local restaurant, donated coupons for free burritos as a reading reward. Small incentives were purchased from the Rock and Roll Hall of Fame and Museum in Cleveland.

Karaoke and Teen Idol competitions added more excitement and energy to the summer. Local game shops came to the branches to hold *Dance Dance Revolution* and *Guitar Hero* tournaments. A wide variety of other non-tech programs were held at all the branches in support of the theme. School of Rock, the Rock-n-Roll over Dead murder mystery program from Highsmith, a Song Writing Workshop, and Hip-Hop Dance workshops were very successful. Craft programs included duct-tape CD cases, melted vinyl record bowls, CD art, and making music buttons, guitar picks, and tie-dye T-shirts.

Unplugged was featured on the library's TeenSpace at www. cuyahogalibrary.org/TeenSpace.aspx and in the *Teen Times* newsletter. An annotated booklist of music-related titles was created in both paper and electronic format on TeenSpace and was linked to the library catalog, databases, and Web sites.

The teens absolutely loved the program theme. Participation was up throughout the system. Summer Reading and the various active programs attracted more high school students than in the past years. The music connection was a winner with both teens and staff. The library hopes to weave the music theme and the successful programs into future SRP activities.

Submitted by Bonnie Demarchi

Young Adult Literature Blogs

A Chair, a Fireplace and a Tea Cozy
  http://yzocaet.blogspot.com
The Goddess of YA Literature (Teri Lesesne)
  http://professornana.livejournal.com
Oracle
  http://barbara-gordon.livejournal.com
Reading YA: Readers' Rants
  http://readersrants.blogspot.com
Swarm of Beasts
  http://swarmofbeasts.blogspot.com

● ● ●

Pass the Book

Teen readers in grades six through twelve participated in a Pass the Book reading program at the 26 Cuyahoga County Public Library Branches near Cleveland, Ohio, from April to August. The program was based around a Web site, Pass the Book, that had project guidelines, links to author interviews and author Web sites, interactive online book discussions, and a method of tracking copies of books. This program was inspired by an article in *School Library Journal*, "Pass It On" (May 2004).

Teens chose the title *Swallowing Stones* by Joyce McDonald in a Teen Read Week survey during the previous fall. A logo contest was held until December, and a graphic design high school student designed the winning Pass the Book logo. The library

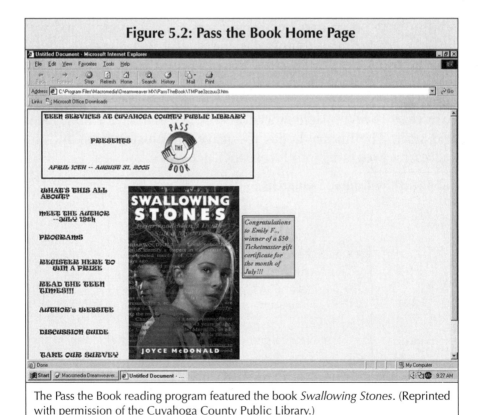

Figure 5.2: Pass the Book Home Page

The Pass the Book reading program featured the book *Swallowing Stones*. (Reprinted with permission of the Cuyahoga County Public Library.)

system originally purchased 600 copies of *Swallowing Stones* and then ordered an additional 100 copies in May to meet the demand. The books were numbered, labeled with instructions, and distributed through the schools. Additional print and audio copies were purchased for the branches, and the circulation was tracked throughout the program. Book discussion sets were provided for programs, complete with discussion questions.

A special staff in-service was held in March for an overview and to plan additional support programs, which included readers' theater, CSI Cuyahoga, a writing workshop for teens, and more. A Web site and READ posters were designed. Online surveys were available at the Teen Space Web site, as well as a link to Joyce McDonald interview at Authors4Teens.com Greenwood database and booklists. The read-alike booklists were used for displays and featured in the *Teen Times Newsletter.*

The program kickoff was held during the National Library Week, with visits to the schools to distribute the books. The teens read the books and noted the book number. They logged into the Web site to track their copy, made comments about the book, and learned more about the author and similar reads. The books traveled as far as Montana and Florida from Ohio. An author visit with Joyce McDonald was arranged in July. The visit included a pizza lunch and book discussion with teens. Several branches participated with related activities:

- Rewrite the ending of *Swallowing Stones* and win a prize
- Where in (name of town) is *Swallowing Stones*? Photos of the book in various locations were posted, and players guessed the location.
- Write a poem or create a piece of art related to *Swallowing Stones*
- Build a stone wall. Teens write their favorite reads on a paper stone to add to the wall.
- Create a Quotation Rock Garden. Write favorite quotes on a paper stone.
- Place your favorite title in a box and select someone else's favorite from the box to read

The cost of the program: $5,000 for books and prizes, $2,000 for support of the author visit, $100 as a prize for the student who designed the winning logo. The marketing department funded the publicity and licensing of the domain for the Web site. Local branch Friends of the Library groups provided additional raffle prizes. The youth services department purchased a variety of paperback books for each branch to use as weekly prizes, and additional audiobook format copies were ordered for the branches. Twenty-six branch teen librarians and five teen public service associates were on board for the program, as well as the graphic design staff for posters and marketing materials. The marketing director and youth services department manager assisted with the visiting author arrangements.

Teen interest was high; they liked the concept, and participation numbers were up substantially from the previous summer.

Figure 5.3: Pass the Book Comments Page

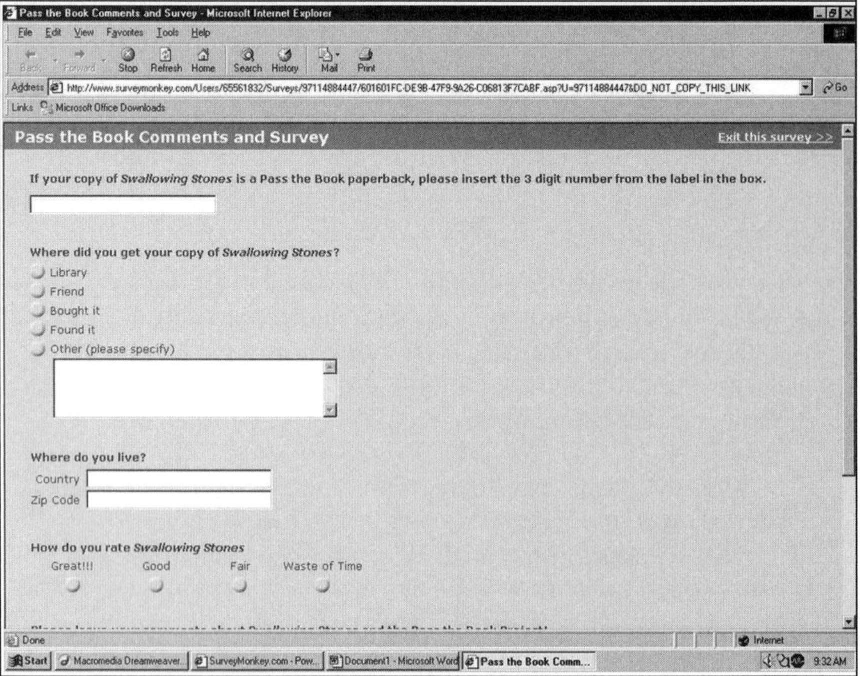

Teens entered information at the Pass the Book Web site to track their copy of *Swallowing Stones*. (Reprinted with permission of the Cuyahoga County Public Library.)

Figure 5.4: Pass the Book Registration Page

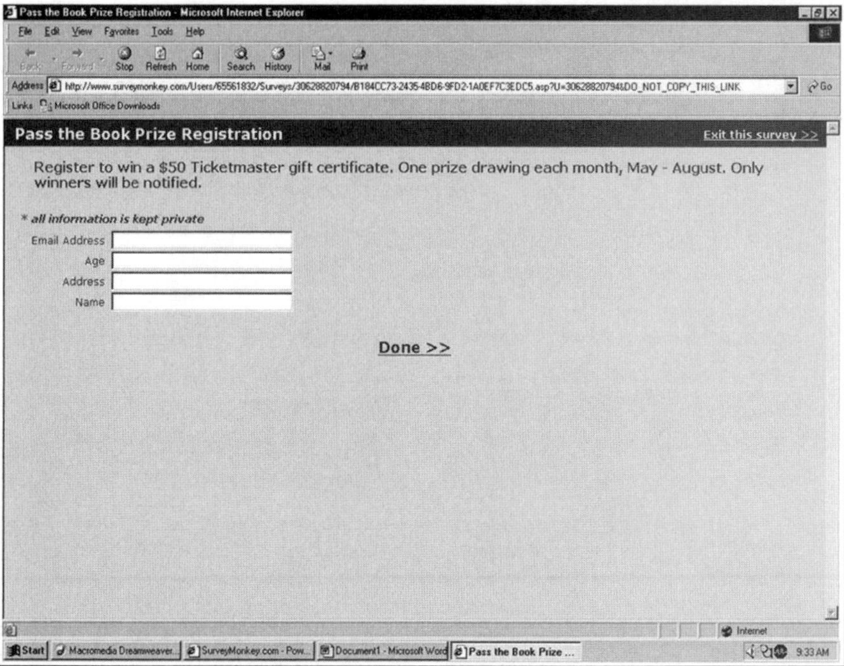

Contact information was collected for a prize drawing. (Reprinted with permission of the Cuyahoga County Public Library.)

Teachers and schools were supportive and involved throughout the program. Circulation of cataloged books was good, and *Swallowing Stones* remained popular the following year. Teens liked the specially designed READ poster, autographed by Joyce McDonald during her visit, which was reprinted and used as a prize.

If Cuyahoga decides to offer Pass It On again, they plan to tie in to the Teen Read Week time period, rather than summer reading, to maximize school participation. They would like to reconfigure their Web site for more accurate tracking and mapping of the book distribution. They also would add a blog, use new technology like podcasting and distance learning connections, and offer online interactive interviews with authors.

Submitted by Mary Arnold

Figure 5.5: Pass the Book Poster

A promotional poster used for the Pass the Book Program. (Reprinted with permission of the Cuyahoga County Public Library.)

✍ Tech Note ✍

Techie Prize Ideas for Teen Reading Programs:

MP3 Players
Digital Cameras
Game Consoles
Nintendo DS
LeapFrog Interactive FLY Pentop Computer
Flash Drives
Gift Cards for Game Stores
Gift Cards for Music Stores
Gift Cards for Video Stores
Gift Cards for Online Stores

Part II

Get Connected
for Education

Chapter 6

The Database, Internet, and Software Connection

Do you find yourself instructing individual teens on the use of databases as needed because they have never heard of them or forgot about them? Teens can also benefit from learning how to use the available software on the library computers. An interesting and active class can engage their attention while you give them new tools to do independent research.

Online Reference Program

Teens ages 11 to 18 attend the Homework Helpers back-to-school program in the fall at the Medina County District Library in Medina, Ohio. The program features online databases and other resources available through the library's Web site. The one and a half hour program is held in the computer area following the outline below. The students find the instruction very helpful and are surprised by all the resources available to them.

The library is building a new computer room for more space and plans to promote the program more to teachers. This program utilizes resources the library already has and educates the teens to be more independent researchers at little cost to the library.

Homework Helpers Program Outline

Intro/Welcome

Basic overview of library Web page at www.mcdl.info

Introduction of Teen Web page

(Cont'd.)

Homework Helpers Program Outline *(Continued)*

Introduction of "Reference Resources"

- EBSCOhost
- eLibrary
- INFOTRAC
- CultureGrams
- Learning Express
- Literature Resource Center
- Opposing Viewpoints Resource Center
- World Book Online
- Knowitnow Presentation at www.knowitnow.org/

Take a tour of Reference Area
- Introduce Reference Department staff

Provide a table with a variety of other helpful homework resources and material

Provide bookmarks with program information

Questions and Conclusion

Submitted by Daphne Silchuk-Ashcraft

Post information and updates about databases on RSS feeds and blogs on the library's Web site.

Research Skills for Seniors

Each September the senior class of St. John Central High School in Bellaire, Ohio, shares one last group effort to enhance research skills that will take them into their college years. The students are immersed in the "Back to the Future" library experience for four hours, with presentations on the best use of all areas of the Mellott Memorial Library building. Although the library remains open to the general public, the entire facility, including some areas typically off-limits, is opened to the St. John students.

In preparation for the visit, the students work closely with the school librarian and the senior English teacher to design the entire trip. They even design a menu for their lunch at Gulla's nearby. Collaboration with the public library via e-mail, telephone, and advance visits is extensive. All topics and areas of interest are collected and forwarded to library staff to facilitate access on the day of the visit.

Practice sessions are conducted in the school library for most seniors, allowing them to anticipate more resources available at the public library and through its linked catalog. The high school library is used as a demonstration library for the students. Although the students can access the public library catalog and databases from home or school, the Reference Section and professional staff are only available on-site. The more examples of research centers a student can experience, the better that student will do when faced with a new research situation in college.

Sister Pat Bartolo, chair of the English Department at Bellaire St. John's High School, and Ed Jepson, the school librarian, assign research projects in mid-August. The effort is collaborative to reinforce all research efforts of previous years' schooling and to prepare students for even more sophisticated efforts in college. The school covers the entire research process and coordinates topics and resources with the staff at the Bellaire Public Library.

The program opens with a helpful virtual tour of the facility, with library maps and photos that can be seen at www.bellaire. lib.oh.us/VirtualTourv3_files/frame.htm. The virtual tour is held in a modern auditorium with a projection screen, Sharp Digital Multimedia Projector, a laptop computer tied into an audio system, and a microphone. Static photos are used to locate and guide students to the sections of the library that would be most helpful to them. A demonstration of the Electronic Databases and Catalogs offered through the facility follows the tour. The library staff is aware of the topics that are being researched and use actual student-generated examples in the presentation.

All of the library resources are used on this trip. The students use the entire facility, including the kitchen, children's section, elevators, copiers, and printers in addition to the actual texts, papers, periodicals, guides, etc. The school provides two people, one teacher and one teacher librarian, and the public library provides two MLS librarians. In practice, the entire library staff becomes involved in helping the students through the process. The restaurant reserves space for everyone, and the students buy their own lunch, which means minimal expense for the library.

Figure 6.1: Back to the Future Senior Class at the Library

Sister Pat Just assists students at work in the lower tech area of the library. This library was designed to take maximum advantage of natural light. (Reprinted with permission of the St. John Central High School.)

Figure 6.2: Back to the Future Senior Researcher

A senior student makes use of a terminal to access items in the Ohiolink databases to work on her research. (Reprinted with permission of the St. John Central High School.)

Figure 6.3: Back to the Future Senior Videographer

Several class members captured the day's events for possible use at school. The entire virtual tour and database exercise were recorded for digital replay on school or home computers for a refresher of the more complicated aspects of searching. (Reprinted with permission of the St. John Central High School.)

The day is filmed and photographed, and students give written feedback so future trips can be improved. Feedback to date has been positive, upbeat, and supportive. The hands-on use of the Bellaire Library and the personal contact with their professional and paraprofessional staff was very helpful, and the students left feeling they had accomplished a great deal in a brief period of time.

Submitted by John Kniesner and Ed Jepson

● ● ●

E-books Scavenger Hunt

Middle and high school students at the Meadows School in Las Vegas, Nevada, celebrated Library Lovers' Month (in February) by Finding eLove in the school library. In previous years, the library had conducted a contest entitled "Find Love in the Library," where students and teachers could find book titles with words like "love," "friendship," "relationship," "romance," and "heart." A winner would be drawn from all the entries each day for a prize of candy. The library had just received a large number of e-books, and students and teachers needed to be shown how to access and use them. It was decided to change the quest to "Find e-Love in the Library" and require contestants to find answers to questions in e-books.

To change the prizes for participation, teens were asked what motivated them. Since uniforms are required at the school, the teens were overwhelmingly in favor of being granted "Free Dress" as a reward, so it was decided that everyone who participated and found correct answers would be eligible for "Free Dress" on Library Lovers' Friday.

Students were directed to use e-books to find answers to questions. They were able to access the e-books on campus and from home. Simple instructions were given about how to access and read an e-book. A volunteer composed questions from 15 of the books, and then eight different forms were made using questions from three books on each. The forms were duplicated on different colors of paper so that it would be easy to see if friends were working together on the same questions. Book titles, questions, and answers were entered in a spreadsheet so the answers could be verified easily.

The activity was announced at an assembly and through the weekly bulletin. Flyers were posted throughout the upper and middle schools. Instruction sheets and entry forms were placed on the charging desk of the library. When the completed forms were collected, corrected, and discussed, students were given individual assistance if necessary. Most were able to follow directions and find correct answers without assistance.

Final entries were collected and corrected, and a list of students who would qualify for "Free Dress" was submitted to the administration. Lists were posted and distributed to teachers. On Library Lovers' Day, students in "Free Dress" were acknowledged informally and engaged in discussions about their newly acquired ability to use technology to access and read e-books.

The library plans to promote the next Find eLove in the Library program more widely and more aggressively. Prizes will be solicited from the business community for teacher participation.

Quotes from Teens

"Now I know how to use e-books."
"I always wondered what an e-book was."
"It's so easy."
"I can read these at home."

Figure 6.4: Find eLove Poster

Are you one of the few people at TMS who actually knows how to find and read an eBook?

If you are, it will be easy for you to qualify for free dress February 24, Library Lovers' Day. All you have to do is find the answers to 3 questions, using the library's eBooks.

If you don't know how to find and read an eBook, you will. Just pick up your entry form and follow the directions.

Posters alerted students to participate in the Find eLove contest to win a "Free Dress" day. (Reprinted with permission of the Meadows School.)

Figure 6.5: Find eLove Directions

Library Lovers' Month—February 2006
Find eLove in the Library—Find eBooks!

Find the answers to three questions using eBooks and qualify for Free Dress, February 24. All entry forms are due February 22, before 1:00 p.m.

To find and use eBooks:
Open STD_US_OPAC
    Type the title of the book in the search term space.
    Find and select the book that shows an "e" in front of the title.
    Select "Available via Gale Virtual Reference Library" to open the book.

If you are asked for a group ID: **nv_tmsch**   If you are asked for a password: **mustang**

(Once you have done this, you shouldn't have to enter ID and password the next time you use the same computer to look for eBooks.)

Take a minute and look at all the things you can do with an eBook (print, email, create a citation, search the index, list of illustrations or table of contents)

Check the "within this publication" box in Quick Search, and you're ready to start looking for answers to the questions.

Step-by-step directions led teens to discover e-books in their school library. (Reprinted with permission of the Meadows School.)

Figure 6.6: Find eLove Questions

Name_____ I♥ebooks!

Find eLove in the Library
Locate *correct* answers to these questions and qualify for Free Dress, February 24.

All forms must be submitted by 1:00 p.m. February 22.

Book	Question	Answer
Acne	What page has the first mention of Accutane?	
Headaches	How many stages are there in a migraine?	
West Nile	When was this book published?	

1

Name_____ I♥ebooks!

Find eLove in the Library
Locate *correct* answers to these questions and qualify for Free Dress, February 24.

All forms must be submitted by 1:00 p.m. February 22.

Book	Question	Answer
Aids	How many chapters are in the eBook on aids?	
Chronic Fatigue Syndrome	On what page is expert Katrina Berne quoted?	
Hunting Terrorists	What 3 countries did Pres. Bush call "The Axis of Evil?"	

2

Teens answered questions using e-books as a resource to win the contest. (Reprinted with permission of the Meadows School.)

Submitted by Ruth Mormon

● ● ●

Classroom Performance System

Students and teachers at the Coshocton County Career Center in Coshocton, Ohio, are enjoying utilizing the Classroom Performance System (CPS) in their classes. The interactive technology has several features that make learning and teaching more fun for all ages:

- Engages all students
- Provides instant feedback to teachers and students
- Collects all objective student performance results
- Increases the flow of student performance data
- Eliminates the administrative tasks of grading objective questions (both homework and tests)
- Provides an electronic grade book with reports

The instructor for Business Marketing Careers junior class has used CPS to study for tests. The students love the hands-on activity. The teacher can quickly see how knowledgeable the students are on the subject because their grades are shown on the system, as well as see questions on which they are weak. Some of the other teachers use the CPS to give tests, as it automatically grades them.

The CPS consists of a wireless response system, radio or infrared, that communicates between handheld response pads and a receiver unit. Teens can answer questions shown on a wall or movie screen anonymously with the numbered handheld pads and can see results immediately on the screen. The software makes it easy to create multiple-choice questions and Jeopardy-type games. Images can be added to each question. The cost of a CPS is based on the number of site licenses needed.

More information about the Classroom Performance System and ordering information is available at: www.pearsonncs.com/cps/

Submitted by Sandy Hess

✍ Tech Note ✍

There are many programming possibilities for using CPS at libraries. Libraries and school classrooms can partner to present a program featuring the CPS by creating book-centered quizzes as part of a classroom visit or library tour. Public libraries can use CPS to put a technology twist on tried-and-true Battle of the Books programs. Gameshow-style teen programs and trivia contests can incorporate the audience's responses with CPS. The system could even be used for staff training.

● ● ●

Database Tutorials

Tech-savvy teens can script and create instructional tutorials to help patrons of all ages use library online resources. The tutorial videos can be loaded directly on to the library's Web site. Tutorials can be created to help patrons use the catalog, databases, and search engines, or even create a page at MySpace. Scripts should cover step-by-step instructions.

Camtasia Studio is software you can use for recording, editing and sharing high-quality screen video on the Web, CD-ROM, and portable media players, including iPod. The software is easy to use. Once the tutorials are recorded, they are exported for the Web, which creates Flash movies that can be uploaded onto the library Web site. A microphone and a computer with the following specifications also are recommended:

- Webserver to host a file size 10MB to 30MB
- Microsoft Windows 2000 or Windows XP
- Microsoft DirectX 9 or later version
- 1 GHz processor minimum
- 2.5 GHz processing capacity(for PowerPoint and camera recordings)
- 500 MB minimum RAM
- 1.0 GB of memory space
- 60 MB of hard-disk space for program installation
- PowerPoint 2000, 2002, 2003, or later version

Camtasia costs $319 for one account, or $1,295 for five users. Microphones can be purchased for under $30. Ordering information is available at: www.techsmith.com/camtasia.asp?CMP= KgoogleCStm

---

Submitted by Jami Schwarzwalder

Chapter 7

The Technology Instruction Connection

Teens learn a lot on their own because they are willing to experiment, but they may not know all the cool tools available to them or how to use them safely. The programs in this chapter are geared to do just that. The wide range of topics and flexibility of these programs makes them adaptable to any library and any group of teens. Little or no cost may be involved with the programs that use the software and hardware the library already owns or has access to. Libraries and schools can also apply for grants at stores such as Best Buy for needed equipment. Take a look at the application at: http://communications.bestbuy.com/communityrelations/docs/StoreDonationApplication.pdf

Social Networking Sessions

Teens come to ImaginOn in Charlotte, North Carolina, every Wednesday afternoon to learn what's new in social networking technology through the Be Smart-Wired program. The series of one and a half hour programs run from September to December and each session features a different way to connect.

The sessions are held in different areas of the library, depending on the technology used. For example, podcasting was held in the library's podcasting booth, and some sessions are held in the computer lab.

The Be Smart-Wired series was inspired by teens who wanted to post photos on their MySpace pages. The result was a Glamour Photo session in the computer area of the library. After taking the photos, the teens learned to use Photoshop to manipulate their pictures and upload them. The teens and the four

staff members helping them were busy for the full 90 minutes of
the program.

Figure 7.1: Be Smart-Wired Did You Know? Handout

Did You Know?

Your Library uses these sites to help you get to
know people and find information:

MySpace
www.myspace.com/libraryloft
Get your ImaginOn. Connect with other teens, authors, and find out about
free programs @ the Loft.

Studio i
www.imaginon.org/studioi.asp
Create your own video, send it to YouTube or even your MySpace page.
Ideas are endless!

The Gaming Zone
http://thegamingzone.blogspot.com/
Check out our latest video gaming programs throughout the library system.

Flickr
www.flickr.com/photos/libraryloft/
Recognize someone? Photos of teen programs at the library.

Ask a Librarian
www.plcmc.org/asklib/default.htm
Stay up late? No problem. Ask a Librarian 24 hours/7 days a week.

Live Homework Help
www.plcmc.org/asklib/default.htm
It's more fun to do homework online. Use your library card, and contact a
tutor.

Reader's Club
www.readersclub.org
Send in your review of a good book! Add your comments to other reviews!

Internet Safety Policy
www.plcmc.org/aboutUs/policiesInternetSafety.htm
Because we care and want you to be safe on the Internet.

300 East 7th Street   www.imaginon.org   (704) 973-2728

A handout directs teens to social networking and safety Web sites. (Reprinted with
permission of Public Library of Charlotte and Mecklenburg County.)

Figure 7.2: Be Smart-Wired Handout

Be Smart-Wired

Get an adrenaline rush and stop by for a free
workshop in the **Tech Central lab**. Interact
online. Create. Impress you friends.

*Every other Wednesday, September–December
4:30 pm–6:30 pm
For Teens, 12–18
Registration requested (704) 973-2728

Sept. 13
Podcast: www.libraryloft.org/podcasts.asp
Be heard! Read a poem, rant, or belt out a song, record it, and put it on
your site (and the library's!).

Sept. 27
MySpace: www.myspace.com
Glamour shots for your page. Be the diva you really are!

Oct. 11
Del.icio.us: http://del.icio.us/
Create an account to collect resources for homework.

Oct. 25
LibraryLoft.: www.libraryloft.org
Contribute to the library's Web site. Write reviews, help maintain pages,
and suggest improvements.

Nov. 8
Second Life: www.teen.secondlife.com
Create an avatar, and be a part of this virtual world.

Nov. 29
Author chat: Ask questions and chat with author Gail Giles about her
writing process and more.

Dec. 13
Library Thing: www.librarything.com
Connect with other people who read the same things as you.

Dec. 27
Share your stuff! Show off your favorite site that you created (MySpace,
blog, wiki, Second Life character, etc.).

300 East 7th Street   www.imaginon.org   (704) 973-2728

A publicity flyer advertised workshops for learning how to use popular Web sites.
(Reprinted with permission of Public Library of Charlotte and Mecklenburg County.)

Another session featured podcasting. The library used Audacity (available as a free download at http://audacity.sourceforge.net/) for podcasting and already owned a microphone. Teens in the library were interviewed in the podcast about the technology they liked to use. Teens also recorded a scavenger hunt in the library. YALSA will use the ImaginOn podcasts to promote Teen Tech Week in March 2007.

Other sessions have featured an online chat interview with young adult author Gail Giles, using del.icio.us for homework, and exploring Teen Second Life: a Virtual World for Teens at http://teen.secondlife.com/. The teens have enjoyed the sessions so far and always want to know and do more. They will have the opportunity to share their work during the December sessions.

Submitted by Kelly Czarnecki

What is del.icio.us?

Del.icio.us is a social bookmarking site where you can bookmark your favorite sites and make them accessible from any computer. You can share your bookmarks with other people or keep them private and browse other people's bookmarks. Available at: http://del.icio.us/

● ● ●

Game Design

On Wednesday afternoons in June and July, teens come to the North Regional Library in Coconut Creek, Florida, for Tech Lab, an hour of instruction on game design. The Teen Advisory Board and the participants in the programs decided what projects they would do at the first "getting to know you" session. Other projects were discussed as well, such as blogging, Web design, podcasting, and better Internet searching skills, but game design was the overwhelming favorite.

The sessions were held in the computer center and a shareware program called Gamemaker (www.gamemaker.nl/) was used. The first month was more formal class instruction where the teens all tried to make the same game and learn how to use the software. The second month involved taking ready-made games and changing certain elements of the games to learn more complex skills. The final days were spent working on their own games, which were generated into standalone games that they could play anywhere. The teens were very proud to create a finished game. Those who started later in the program had some difficulties finishing their games.

Library books, magazines, and databases that could help them with game design in class were promoted. The library also promoted the summer reading program during class. There were usually two to three staff members present at each session. The Friends of the Library paid $170 for a 20-seat license for the shareware.

Next summer, the library will be offering a larger variety of software so that it will be possible to make more elaborate 3D graphics. The library has received a grant to expand the program to other libraries within the system. The future programs will be more self-directed and have less of a classroom feel.

Submitted by Katherine Makens

● ● ●

Teen Tech Camp

Teens, ages 13 to 15, meet for six hours a day for five consecutive July days to connect with emerging technology at the Memphis Public Library and Information Center in Memphis, Tennessee. The library utilizes the Central Library Computer Training Lab and the Training Room, and the demonstrations are conducted in the library's WYPL TV and Radio studios.

The first camp developed the library's teen Web page at www.memphislibrary.org/humanities/TEENPAGE.htm. They used HTML Kit and HTML Tidyplugin software for it. Teens met with real mentors in real time for the second week-long technology camp. The second camp produced Web cam projects available at www.youtube.com/results?search_query=memphis+public+library. They used Web cams, video editing, digital photography, flash drives, desktop and laptop computers, Microsoft Movie Maker software, and PowerPoint.

Figure 7.3: Teen Tech Camp Logo

The Teen Tech Camp logo branded the program and looked great on T-shirts. (Reprinted with permission of the Memphis Public Library and Information Center.)

The library and teens developed a Web blog available through the library's teen Web page so teens could correspond with each other and the library on an ongoing basis. Teens were encouraged to blog about technology issues, and librarians communicated about upcoming programs. The blog was short-lived because conversation was only from the library to the teens, with virtually no feedback from the latter.

At the conclusion of the first Teen Tech Camp, participants were asked for their input and suggestions. They requested more video-type projects, which led to the establishment of a Web cam project for the next year. The teen blog also was used to assess interests. A participant from the first Teen Tech Camp assisted as a mentor for the next camp.

The Memphis Chapter of the Society for Information Management (SIM) provided all of the funding for the Teen Tech Camp. In addition to financial support, they offered their time and support from member companies to come show teens the latest in technology products. Local restaurants donated lunches for the campers.

Planning began after confirmation of the partnership with the Society for Information Management. The goal was to provide programming to connect young teens with emerging technology. In today's information age, SIM and the library hoped these experiences would encourage youth to understand that the many facets of technology offer a wide range of study areas and skill levels to pursue as career options. The Central Library "Teen Team," librarians who work with YA customers in the subject departments, met with the assistant director for Library Advancement, the Youth Services coordinator, and the IT assistant director to develop themes, titles for the programs, and curriculum outlines. Potential instructors and the technology necessary for the program were identified.

The application process was developed to reach as broad a pool of potential participants as possible. Applications were available at all Memphis library facilities and were sent to every middle school in the Memphis City School System, the Shelby County School System, as well as private, independent, and home school associations. They were also downloadable from the Memphis Public Library and Information Center (MPLIC) Web site for several weeks before the academic spring break. As part of the application process, each potential camper had to submit a recommendation from a teacher. In the second year of the camp, three schools came forward to promote the camp to their students. Over 40 youth applied each year.

Following a blind review by Teen Team members, participants and alternates were chosen. The groups were remarkably balanced regarding race and gender. Letters were sent out to every applicant with information on their application status. Two weeks before the camp began parents and teens were required to attend an orientation meeting at the Central Library. All parental permissions/releases were signed, and emergency contact information was verified.

Each morning of the camp week, members of the Society for Information Management presented a short program, employing

PowerPoint and hands-on experience with the newest and most desirable technology on the market. SIM named this info sharing, "Bright Shiny Objects (BSOs)." Following this, campers received instruction on the various technologies that would be employed during the camp, worked on their chosen part of the projects, with breaks for snacks, lunch, and stretching.

The camps culminated in demonstrations for friends, family, library staff, school administrators, and the media. On the final day, each camper was awarded a certificate of participation, a specific award for their particular talent and contribution to the program, a T-shirt with the Teen Tech Camp logo, a goodie basket to remember the camp experience, a Web cam, and a 1GB flash drive containing their finished technology projects. As part of the closing exercises, all campers evaluated their experiences and offered requests for upcoming technology for programming and future camp sessions. Staff also evaluated the camp in detail within a week of the end of the sessions.

Numerous library resources were used for the camp including books, Tennessee Electronic Library databases, software, and programs. The MPLIC television and radio stations were used as resources for the individual projects. The curriculum also included other activities and instruction to supplement the technology. The campers found their own library resources to highlight in their various technology projects. Both the teen Web site development and podcasting projects were based on camper-chosen library resources.

The Central Library Teen Team includes representatives from four departments at the Central Library (Sciences, Humanities, History, and Library Information Center). The co-chairs responsible for executing the project were chosen from the Teen Team: Sciences Department's John Lloyd and the Humanities Department's Clare Coffey. In 2006, additional assistance included one adult volunteer "activity coordinator," one teen volunteer, and at least one presenter from SIM each day. Behind the scenes, IT staff helped with technology advice and assistance, and the Development staff solicited donations and corresponded with donors.

The expenses for the program were as follows:

T-shirts	$ 350
Flash Drives	$ 450
Webcams	$ 475
Headsets	$ 150
Sub-total	**$1,425**
Printing and Postage	$ 400
Supplies	$ 50
Snacks/Food	$ 375
Prizes and Recognition	$ 125
Paid Assistant	$ 200
Reference Resources	$ 100
TOTAL	**$2,675**

Figure 7.4: Teen Tech Camp Group Photo

Tech Campers gather for a photo. (Reprinted with permission of the Memphis Public Library and Information Center.)

The 2005 group was largely composed of 15-year-olds who were interested in career-based internships. They enjoyed the experience of filming a 30-second commercial for the library to add to the Web page they designed. The 2006 group had a majority of 13-year-olds who liked using computers so much that they wanted the camp to continue beyond the one week scheduled. Each year, all teens spoke highly of their experiences.

Future camps will provide this experience to more teens by offering two sessions. As more teens are added, staffing needs must be included in those considerations, such as hiring additional instructors or recruiting volunteers with specific skills.

Submitted by Betty Anne Wilson, Clare Coffey, and John Lloyd

✍ What Is a Flash Drive? ✍

A flash drive is a small device used to transport and back up data and is also capable of holding software applications. The flash drive fits into a USB (Universal Serial Bus) port. They are smaller and more reliable than floppy disks and store much more information. When purchasing computers for public use, look for USB ports in the front of the hard drive for easy patron access. Libraries can sell drives to patrons with library logos on them or give them away as prizes for summer reading. Currently, flash drives can be purchased for about $8 and up, depending on the amount of memory.

● ● ●

Research Papers and More

High school students working on research papers get after-school help at the Evansville Vanderburgh Public Library in Evansville, Indiana, every other Tuesday, September through November, from 3:30 to 5:00 pm. The teens gather in the Tech Center at the library system's Central Library. The Tech Center is a private computer lab with teaching equipment, a teacher's monitor and large-screen display, and individual computers for up to 16 students.

This program was designed to teach teens about the technology within their own PCs. It included six different sessions, each with a different tech topic pertinent to teens. Tech Teens evolved after the teen advisory board requested more technology and computer classes for the fall program plans. They specified two of the topics and gave input about formulating the other topics.

This program did not require any specific collaboration other than working with the local schools to let them know about the opportunities within this program. Teachers were pleased with the topics of the classes and readily promoted the program to their students. Some teachers even requested reservations for themselves because they admitted they did not know the information being taught.

Plans for the program began early in the summer, when a librarian met with the teen advisory board to discuss ideas for fall programs and the teens requested more technology training. Topics and format, including days and times, were discussed. Topics for the program were selected in June. Two other librarians and staff members were recruited to teach a few of the classes, owing to their subject expertise. The Teen Services librarian taught three of the classes. The Library's Webmaster taught two classes, and the head of the library's Reference Department taught one. Publicity was submitted to the library's Graphic Design Department in July for the overall teen program brochure and for posters and bookmarks about the Tech Teens program. In August, the Teen Zone bulletin board display featured the program for a month before the first session, and notices about the classes were mailed to the area's middle and high schools. Information about the programs was also given to the local home school listserv. Classes were featured on the library's homepage as a "What's Happening" spot: they were also posted on the Central Library's monitors at various locations throughout the library. Posters and bookmarks were displayed at the Central Library and the library's branch locations. Registration opened and the classes started in September.

Week 1: "Date-a-What??" Teens learned to use databases to search "below the surface" of the Web and find the really good answers. This program included training on using the library's databases, as well as tips for searching within those databases.

Week 2: "Research Your Research Options." Google's good, but it's not the only option. Teens used other tools for locating information on the Web. This class gave participants a list of other search engines, all of which were new options to the teens, as well as information about subject guides. Teens were amazed at the differences in results from the various search engines as the instructors did sample searches for comparison.

The classes continued through October.

Week 3: "Become a PhotoShop 'Phool'!" Teens got the "phacts" for doing "phantastic" photo editing. This topic was re-quested by the teens on the Teen Action Group, the Library's teen advisory board.

Week 4: "HTML: It's Not a Secret Code." Hypertext Mark-up Language: the language of Web sites, word processors, and more. Teens got to learn it, write it, and use it. This session was another topic that was specifically requested by the teen advisory board.

The last classes were held in November.

Week 5: "The Internet Knows All...Or Does It??" Teens learned to avoid misinformation, poor answers, and deceptive "facts" by learning how to evaluate Web sites. This class was cre-ated based on the information in the book by Frances Jacobson Harris, *I Found It on the Internet.* The class began with a true story about a local teacher who gave his class an assignment on the "Blue Peruvian Frog Plague" and flunked all of the students' research papers when they quoted a Web site that the teacher had created as a ploy with false information that fooled all of the stu-dents, because they did not know how to evaluate the site.

Week 6: "Beyond Googling." Teens learned how to maximize their Google searches like professionals. Students learned how to use various codes and options within Google to streamline their searches, thereby make them more effective.

Expenses for the series were negligible. Some of the classes had photocopied handouts, and there was the standard publicity from the library's Graphic Design department. Cost to the teen programming budget was zero.

One teen commented after the database class, "Wow! This was really interesting!" She added that the information would be "really helpful" in working on her senior research paper. After the search engine comparisons, a teen thanked the librarian for showing him more search options. He said he was not aware there were so many other search engines, and he was amazed at the information available on the various subject guides viewed in the class. Here is a quote from his e-mail: "It seems like the Internet compares to an iceberg in that only 10 percent is visible." He was pleased to learn how to dig deeper than that 10 percent level. The remaining classes were filled to capacity.

The teachers are enthusiastic about promoting this program to their students, and the teens are excited about learning and appreciate what they're learning. This program proved to be a viable, needed, and well-attended one for teens and was also easy on the budget.

Figure 7.5: Tech Teens Class Photo

The search engine class was most popular with boys! (Reprinted with permission of the Evansville Vanderburgh Public Library. Photograph taken by Josh Weiland.)

Submitted by Maryann Mori

● ● ●

Teen Tech Group

The Teen Techies, formerly known as the Teen Webbies, get together every week at the Homer Township Public Library District
meeting room in Homer Glen, Illinois. The first few meetings
were spent discussing goals and making a timeline. They decided
their first project would be creating a teen Web page.

Teens viewed other library teen pages and took notes on ideas
and placed them on their Wiki. Ideas were voted on to decide
which elements would be included in their new Web page. The
group then had a few short HTML lessons, and the teens continued to update the site and incorporate new technologies. They
have learned HTML and blogging and placed reviews on the
Teen Reviews blog. Additionally, they have evaluated content for
the del.icio.us site and will be podcasting this fall.

Participating teens love this program, and the only expenses
are refreshments. The Teen Techies are very proud of their Web
page. They have all agreed to make this a year long program.

Submitted by Alexandra Tyle

✍ What Are Blogs and Wikis? ✍

Blogs and Wikis are two popular ways to share information on the
Web 2.0 Internet, but which would you prefer? Both platforms are
free. Blogs are written by one or more persons who post articles or
news on any topic in reverse chronological order, like a journal.
The author may or may not allow the reader to comment on each
entry. Blogger, at www.blogger.com, is a site where you can create
a blog, add photos and links, and customize the appearance to
suit the topic or your personality. There are step-by-step instructions, and additions and revisions can be made as desired by the
author. The blog can be sent via FTP to the library's server. Frequent posts will keep your readers coming back to your blog.

A Wiki is more conducive to a sharing of the collective intel-
ligence. Many people can become members of a Wiki and edit
and add information on one or many topics. The most famous
Wiki is Wikipedia.org, an ever-growing and ever-changing collection of knowledge and information authored by thousands

of people. Pages are linked together by keyword links throughout the text to provide more information on any topic mentioned. There are Wiki sites such as www.wikidot.com where you can create your own Wiki. Wikidot has features like a discussion forum and tags to organize the information. The access to your Wikis can be as private or public as you desire.

Wikis and blogs can be used to write about movies, books, Web sites, and games, or to create an online book club.

● ● ●

Business Card Workshop

The Teen Advisory Board of the Greenville County Library System in Greenville, South Carolina, asked for more programs in the computer training lab. After researching what other libraries were offering, particularly the Austin Public Library, the library in Greenville initiated the Teen Tech Thursdays series during the summer reading program.

One of the workshops taught teens how to create their own business cards using Microsoft Publisher. Participation was limited to 10 teens since there was only one color printer available. After the 10 teens registered, there was a waiting list and many calls asking for the program to be offered again. Two staff people were needed, as the color printer was in another area of the library and the extra person was a runner when the printing started. The program was 90 minutes long, with one hour to create the cards and a half an hour to print.

The teens were invited into the training lab and asked if they had any experience with Microsoft Publisher. All participants were familiar with the software but had never created a business card. The librarian discussed what information should be included on a business card, and the participants shared thoughts about the kinds of businesses they wanted to advertise. Sample business cards were shown on a projection screen, followed by a demonstration on how to get started. The librarians circulated, answering questions and giving assistance when needed. Teens were reminded periodically when the printing would begin.

The cost was $10 for 1,000 business cards. Each teen printed two sheets for 20 cards. The remaining cards will be used when the program is repeated. Be sure to have enough color ink, as their cards were very colorful with very little white space. Test-print the sample cards to know where to set the margins, since some printers do not print exactly what you see on the screen.

The teens had a lot of fun and were very serious about creating their cards. Most cards were for babysitting and lawn mowing, but one boy made cards for his mom's new home business and another made some that stated he was a video game tester.

Submitted by Kristin Whitworth

● ● ●

Game Design with a Corporate Partner

Teens ages 12 to 18 come to the Minneapolis Public Library (MPL) Teen Central in Minneapolis, Minnesota, for an ongoing series to learn how to create video games. The longer sessions run two to four hours and shorter sessions are offered as Open Gaming Labs. The Open Gaming Labs are for teens who have already learned the basics of the software to work on their video games. The longer sessions, offered in the summer and fall of 2006, were more in-depth, teaching the many aspects of the software program in addition to offering time to work on a video game.

The programs are held in the Best Buy Technology Center at the Minneapolis Public Library's new Central Library. Scratch, a software program developed by MIT Media Labs to create your own video games or animated sequences, is used for the workshops. Students use desktop computers to work on their projects. Other technology workshops complement this program and allow teens to learn skills that can help them enhance the video games they create. The additional sessions focus on using Flash, Crickets (motion-controlled devices that can be used to create digital art and more), and the next phase, AnimeStudio, which can be used to create animated video.

Gaming Studio teaches teens how to use Scratch to create their own video games or animated sequences. During the first season, sessions lasted three hours; teens could attend one session or more. Teens often attended every session so that they could continue working on their games. Sessions were offered three to four times per month for several months.

Each session begins with an introduction to Scratch and some guidance on using its basic features. Teens then are given time to practice and work independently on a game of their choice. Midway through the program, the group reconvenes, and instructors share new tips or features that will allow participants to enhance their games and give them more sophisticated elements. Participants use primarily computers and the Internet (to find images and patterns for their video games) but also are directed to print materials in the library that tie into their work and interest in video games.

Informal feedback for the program is sought from the Teen Central Teen Advisory Group. After a summer of successful sessions of Gaming Studio, 12 of the most interested, repeat teen participants are invited to attend three intensive training sessions with library and Science Museum of Minnesota (SMM) staff that prepare them to lead their own sessions, the second phase of Gaming Studio. These teens have formed the Teen Tech Squad.

The Open Gaming Labs are held a bit differently than the initial training sessions. They are more informal and last two hours. The Squad gives a basic introduction to Scratch and then works with each participant individually to teach new skills and techniques to enhance the game. All created games are stored on a server and are linked to the library's Web site. This allows showcasing of the teens' work and gives the teens access to each others' games. The Squad works in teams of three to host the Open Gaming Labs and Scratch sessions, where they are paid $25 for each two-hour session to teach adults, children, and other teens how to use the software program to create animation or video games.

The program is part of a large series of Teen Technology Workshops being offered at MPL and funded through a generous grant from Best Buy. The program itself is the result of a valuable partnership between MPL and the Science Museum of Minnesota's Learning Technologies Center. Science Museum staff have served as instructors and facilitators of Gaming Studio sessions and helped train the Teen Tech Squad. They have shadowed Teen Tech Squad members as they offered their first Gaming Studio sessions. Teen Tech Squad members will have access to follow-up training with the Science Museum to build on their knowledge so that they remain the experts in using Scratch and continue to be an asset to teens learning the software. Tech Squad members spend 30 minutes prior to and following each session to plan and evaluate their sessions. They also complete a thorough evaluation as a condition of their being paid.

The instructors from the Science Museum of Minnesota teach this course with a youth development framework in mind. They incorporate elements that help each group build a sense of cohesion, where teens can share their work, receive feedback, and build a sense of accomplishment through the work they have completed. This focus on a youth development framework for teaching aligns the goals of the program with the Minneapolis Public Library's broader approach to teen services and programs.

The third phase of the program will be introduced during the summer of 2007, when the Tech Squad offers sessions focused on creating animation. For this they will teach and host labs using AnimeStudio, a software program that can be used to create digital animation. Also on the horizon is MangaStudio, which would allow them to host Create Your Own Manga workshops.

Instructors for the Science Museum are paid on an hourly basis. The cost for the initial three-hour programs ran approximately $300, which included both SMM staff and other materials that had to be purchased. Teen Tech Squad sessions cost roughly the same, even though they were shorter. A team of three Tech Squad members were each paid $25 for each session and, in these early stages, one SMM staff person still needed to be hired to shadow and give them feedback on their sessions. Eventually, this SMM staff person will not be needed.

This is by far one of the most popular and successful programs the library has offered. The first season had long waiting lists for each program. Teen participants attended programs multiple times to continue working on their games. Feedback about the program from all the teens has been very positive. Teens are interested in creating more sophisticated images and visuals to incorporate into their programs, so future sessions may begin with some complementary programs on 3D animation.

Submitted by Christy Mulligan

Further Reading on Technology Instruction

Richardson, Will. 2006. *Blogs, Wikis, Podcasts, and Other Powerful Web Tools for Classrooms.* Thousand Oaks, CA: Corwin Press.

Chapter 8

Get Connected with Special Audiences

Technology can be an excellent tool for connecting with special audiences. These programs are creative examples of meeting the needs of those audiences with newer technologies. Earphone English utilizes audiobooks in all formats to help ESL students and other students with language difficulties. Teens WAVE @ Central guides young adults in job searching using a local Web site and includes instruction on filling out job applications.

Audiobooks for ESL Students

Teens learning English meet one lunch period per week during the school year at Berkeley High School in Berkeley, California. The program was built in 2000 from the ground up by five teens learning the English language. Each year, new members of the club have made adjustments. From the beginning, Earphone English has been a collaboration among Berkeley Public Library's Teen Services, Berkeley High School's ELD Program, and later, the King Middle School's library.

Participants select and discuss audiobooks from a comprehensive collection on tape, CD, and MP3 files selected for this purpose with regular input from participants. The teens listen to the books on their own time. The Berkeley Public Library (BPL) supplies the materials, including players that can be borrowed by those who need them.

At the middle school, the program is incorporated into the learner's English class. A BPL Teen Services librarian booktalks and leads occasional small group discussions during class times in the school library. At both schools, the library staff has made a concerted effort to have print components of the audiobooks. About 140 students participate through each year, but generally discussion groups include about 10 to 12 students, a BPL librarian, and occasionally an ELL teacher.

The library has built up an Earphone English specific collection of audiobooks in all formats, players, print books, and some craft materials. Some field trips to the public library include watching movies from the public library's collection. (The library has a movie license.) Occasional pizza parties and craft events are also part of the program. A collection of games has been added for participants not immediately engaged in discussion to play while waiting to share.

Two public library librarians each spend about two and a half hours on the project weekly. Occasionally, an Earphone English graduate volunteers with the high school group. Teachers assist by supplying space, commitment, and some discussion interaction at the high school and some listening time at the middle school.

During the second through fourth years, a $15,000 annual grant was used to build up the collection and pay a substitute librarian to give staff release time. The program now costs about $3,000 a year for materials and food and is a part of Berkeley Public Library's regular budget, a component of the Teen Services budget.

Some of the students loved it and promoted it so heavily that, at the beginning of this year, 36 students showed up at the first session. Even students who do not like audiobooks come to the program because they like the community. Some do not want to share in discussions but want to do the listening, which is possible because they can borrow materials from the teacher's closet where extras are stored. Each spring, participants complete brief written surveys about their favorite and least favorite aspects of the program.

Everyone in the program has the special need for English language access. There have been a number of students with physical and developmental issues, all of which could be addressed in the context of the group meeting in public school classrooms.

Submitted by Francisca Goldsmith

• • •

Job Applications Tutorial

The Virginia Beach Central Library in Virginia Beach, Virginia, held a two-hour program in their computer lab to teach teens how to use the WAVE, Web Application for Virginia Beach Employment. The WAVE is a Web site used to apply for jobs within the city as well as helping the applicants word their job applications properly. It was the Central Library Teen Council's suggestion to have a program like this so that they would feel more comfortable with the new online application process for municipal employment opportunities.

Each teen sat at a computer, and the instructor used a laptop and multimedia projector to show them how to get to the Web site and create a user name and password. All the steps of filling out the application and saving it were covered. The teens can use this application to apply for any City of Virginia Beach job. They

Figure 8.1: Teens WAVE @ Central Participant

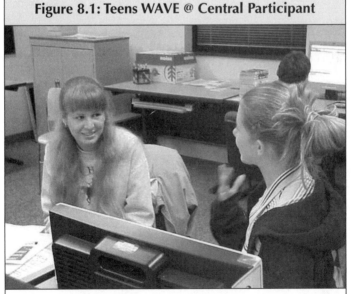

A Teens WAVE @ Central participant learns how to fill out a job application. (Reprinted with permission of the Virginia Beach Central Library.)

were also shown how to edit and copy and paste from other programs instead of filling in the application from scratch if they had a resume on Word or another program. The city official responsible for pooling the applications for employment with the City of Virginia Beach came to talk to the teens about what knowledge, skills, and abilities are sought after in an application. Finally, the teens created resumes on their own, and the librarians helped and answered any questions. The teens were given a booklist to help them create a resume and cover letter.

The teens really enjoyed the program. One of the Central Library Teen Council members is now a City of Virginia Beach employee, and she said the training really helped her with the application. The library plans on continuing this successful program, making any adjustments the teens might suggest. There are no expenses associated with the program.

Submitted by Kelly Greenfield and Susan Paddock

Part III

Get Connected with the Teen Advisory Group

Chapter 9

The TAG Connection

Teen Advisory Groups (TAGs) have proven to be key to developing teen-friendly services and programs in libraries. Some teen groups have become specialized, focusing on just one area of interest and incorporating club-type activities. The groups in this chapter are examples of how Teen Advisory Groups can learn about many of the technologies available to them in libraries and how librarians can help provide teen-friendly, relevant programs.

Teen Council MySpace Page

In celebration of the Teen Read Week 2006, the Pierce County Library System (PCLS) Graham Branch in East Graham, Washington, sponsored a library night for teens ages 11 to 18, complete with *Dance Dance Revolution* consoles and pizza. At this event, the librarian gathered a list of names, e-mails, and phone numbers and distributed handbills promoting the first official meeting of the MyLibrary Teen Council.

Plans to create a MySpace page began at the second meeting and continue to evolve. The new council meets no less than once a month to give input on teen programs, materials, spaces, and issues in the library. The teen council seeks to include a larger membership by providing opportunities for participation online 24/7, including through a page at the social networking site MySpace. At meetings, the Teen Council spends half of the time conducting regular business; the other half is dedicated to creating and maintaining a profile and presence online. The librarian keeps the password and initially controlled the mouse and computer. The laptop is hooked up to a presentation projector, so everyone at the meeting can see what is going on and participate.

Many features of the MySpace profile can be tailored to a teen council's interests:

- List all meeting times and upcoming library events, and/or local concerts or community events of teen interest on the calendar
- Host songs from local bands or book-themed bands like Harry and the Potters
- Blog meeting minutes and collaboratively written reports from library programs
- Teens who miss a meeting, cannot attend regularly, or who do not want to sit through a meeting can participate through blog comments and page comments
- MySpace's groups feature supports other teen groups that meet regularly at the library
- Profile pictures can rotate through pictures of the teen area, collection, and teen-designed group logos
- A podcast can be posted to download, made by group members at the monthly meetings
- An RSS link updates teens when the profile is updated and when a new podcast is uploaded

The goals for this group include:

- Encourage and expand teen participation in shaping library services and collections through use of both traditional (meetings) and modern (Internet/virtual) forms of communication
- Integrate collaborative learning/teaching of technologies for teens at all knowledge/socioeconomic levels into every monthly meeting
- Develop the young adult customer's awareness of, and investment in, the library and larger community

This program supports new and existing teen literacies, combining traditional teen programming with current communication/social networking technologies that many teens already use every day, thus validating and empowering these forms of communication and participation. The program fulfills, in part, PCLS's Youth Services Balanced Scorecard objectives—"develop a customer focus" (teens are the customer) and "deliver contemporary programs and services"—by specifically targeting

young adults through the use of current technology important to them.
The technology and supplies for this program include:

- Laptop with wireless card and Web browser, $3,000
- USB Microphone, $30–$75
- External computer speakers, $100–$150
- Audio editing software
- Digital camera, $300
- Photo editing software
- E-mail application/access
- i-Tunes software
- Computer projector, $500–$1,000
- Projector screen or blank wall
- RSS feeder and reader
- An additional $250 is budgeted for snacks for one year

The MyLibrary Council was promoted with flyers posted in the library and at local schools, an advertisement and link on the Library System's teenspace Web page, handbills given out during booktalks at local schools, announcements at Friends of the Library meetings, and through daily interaction with teens.

Submitted by Michelle Angell

✍ What Is MySpace? ✍

MySpace is a popular social networking site for teens. Catering to teens' desire to socialize while multitasking, users can chat, watch videos, listen to music, watch videos, write, create content, and visit friends' pages. Every MySpace page can be unique to the user. Host a MySpace workshop in which teens decorate their spaces using templates they build themselves or from online resources.

● ● ●

Pen-Pal Podcast

The Carnegie Library of Pittsburgh (CLP) Teen Advisory Council (TAC) of Pittsburgh, Pennsylvania, spent the last half-hour of a meeting creating a podcast to introduce themselves to the Teen Advisory Board of Kansas City Public Library—West Trails Branch in Kansas City, Missouri. The teens planned the podcast content, and the librarian planned the framework for the program.

Any interested TAC members were invited to stay after the meeting to produce the podcast. The teens and the librarian discussed the podcast script and wrote down a rough outline for the audio presentation. The teens nominated the librarian to facilitate the podcast as an interviewer. The laptop and microphone were set up in the middle of the table, and the podcast was recorded using Audacity software. The podcast introduced the group of teens, their advisory group accomplishments, and other information, such as favorite reads. Lots of laughs were involved, which increased the value of the content as the essence of the teen group was documented. After the program was over, a summer teen volunteer helped the librarian edit the podcast. Audacity was also used to edit the volume of teens laughing, in order to establish equilibrium.

An Ourmedia account was created for the CLP teens and the podcast was uploaded there. Afterwards, the podcast was embedded into the CLP Teens blog and the Trails West Teen Advisory leader, Amanda Rodriguez, was notified so that her teen groups could respond to the teen pen-pal podcast. Their group of library advisory teens responded by publishing their own podcast.

✍ What Is OurMedia? ✍

Ourmedia.org is a free Web site where you can publish and store video, audio, text, and image files that you create. Ourmedia allows members to post comments and blog.

This podcast pen-pal program gives teens a chance to use podcasting technology before taking on a big project, such as an ongoing teen radio podcast. Clips of the podcast could be used to promote the Teen Advisory Board in a board meeting or school visit. The program did not cost anything since the library already owned the laptop and microphone.

The CLP Teens Blog entry about the podcast can be viewed at: http://clpteens.blogspot.com/2006/05/listen-to-our-podcast.html,

and the six-minute Our Media Podcast Link is located at: www.
ourmedia.org/node/228166.

Submitted by Stephanie Iser

● ● ●

Gaming Group

Teens in grades six through twelve meet one hour per month to
plan and run gaming events at the Schaumburg Township District Library in Schaumburg, Illinois. At the first meeting, two
Xbox consoles with a *Star Wars Battle* game were set up as a
system link, an XBox 360 with *John Madden Football*, a Playstation 2 with *Dance Dance Revolution*, and a Playstation 2 with
Guitar Hero were available for the first 15 minutes. The library
already owned the equipment from summer programs.

Two staff members, Joe Torres and Jason Larson, ran the
meeting, with the help of a high school senior community volunteer. Joe asked the teens questions about what games the library
should buy for teens and for younger children. They discussed
upcoming new platforms and games and decided on future club
topics, such as trying to design their own video games. Snacks
were available.

The teens who attend these meetings want to be able to play
as a group on system links. While they do give topic ideas, what
they really want is a chance to try different platforms and play
games during free time. For the first few meetings, the facilitators
decided to introduce new games and just let the teens play. In the
future, the group will run programs with mini tournaments for
teens and younger children during spring break.

The gamers can now interact through the library's MySpace
site at www.myspace.com/stdlteen, where they give input on
speakers and topics for future meetings.

Submitted by Amy Alessio

Questionnaire for
Get Connected: Tech Programs for Teens

I f you have recently hosted a successful technology-oriented program for the teens in your community and would like to have information about it published in the upcoming book *Get Connected: Tech Programs for Teens*, please fill out and return the survey below.

Part 1: Your Program
Please type your responses directly into this document.
1. What is the title of your program?
2. Who was your intended audience?
3. When did your program take place (date, time, or occasion)?
4. How long was your program?
5. Where was your program held? Was there a virtual component? If so, please describe.
6. What kinds of technology did you feature in your program?
7. What type of teen participation was there for planning and executing the program?
8. What partnerships or collaborations with local businesses, community organizations, etc., was involved with the program?
9. Describe your program, including the sequence of events.
10. What library resources (books, databases, periodicals, etc.) did you use in your program?
11. How many staff members, volunteers, hired instructors, or presenters did you use?

12. What were your expenses? How did you fund your programs?
13. Do you have photos, logos, screen shots, or handouts from your program? All images must be print quality (300 dpi or better) and saved as .jpg or .gif files. Print documents should be sent as .pdf files.
14. What did the teens have to say about your program?
15. If applicable, what accommodations did you make for teens with special needs?
16. What, if anything, would you do differently if/when you repeat the program?

Part 2: Your Contact Information

This information will be used to contact you if clarifications are needed and also to give you credit for your contribution to the book:

1. Your name:
2. Your job title:
3. Your work e-mail address:
4. Your library:
5. Your library's address:
6. Your library's phone number:
7. Your library's Web address:
8. Your library's blog:
9. Your library's MySpace (or similar) page:

Please send your responses via email to rhonnold@yahoo. com. Be sure to put "Teen Tech Program" in the subject heading. Any accompanying files of images, logos, etc., should be named after the name of your program and your library. For example, if your program was a DDR Marathon and you had a photo to share, name the file of the photo "DDRMarathon_XYZLibrary. jpg." Thank you for sharing your work to celebrate YALSA's 50th Anniversary!

Access for Children and Young Adults to Nonprint Materials

An Interpretation of the Library Bill of Rights

Library collections of nonprint materials raise a number of intellectual freedom issues, especially regarding minors. Article V of the *Library Bill of Rights* states, "A person's right to use a library should not be denied or abridged because of origin, age, background, or views."

The American Library Association's principles protect minors' access to sound, images, data, games, software, and other content in all formats, such as tapes, CDs, DVDs, music CDs, computer games, software, databases, and other emerging technologies. ALA's *Free Access to Libraries for Minors: An Interpretation of the Library Bill of Rights* states:

> ...The "right to use a library" includes free access to, and unrestricted use of, all the services, materials, and facilities the library has to offer. Every restriction on access to, and use of, library resources, based solely on the chronological age, educational level, literacy skills, or legal emancipation of users violates Article V.
>
> ...[P]arents—and only parents—have the right and responsibility to restrict access of their children—and only their children—to library resources. Parents who do not want their children to have access to certain library services, materials, or facilities should so advise their children. Librarians and library governing bodies cannot assume the role of parents or the functions of parental authority in the private relationship between parent and child.

Lack of access to information can be harmful to minors. Librarians and library governing bodies have a public and professional obligation to ensure that all members of the community they serve have free, equal, and equitable access to the entire range of library resources regardless of content, approach, format, or amount of detail. This principle of library service applies equally to all users, minors as well as adults. Librarians and library governing bodies must uphold this principle in order to provide adequate and effective service to minors.

Policies that set minimum age limits for access to any nonprint materials or information technology, with or without parental permission, abridge library use for minors. Age limits based on the cost of the materials are also unacceptable. Librarians, when dealing with minors, should apply the same standards to circulation of nonprint materials as are applied to books and other print materials except when directly and specifically prohibited by law.

Recognizing that librarians cannot act *in loco parentis*, ALA acknowledges and supports the exercise by parents of their responsibility to guide their own children's reading and viewing. Libraries should provide published reviews and/or reference works that contain information about the content, subject matter, and recommended audiences for nonprint materials. These resources will assist parents in guiding their children without implicating the library in censorship.

In some cases, commercial content ratings, such as the Motion Picture Association of America (MPAA) movie ratings, might appear on the packaging or promotional materials provided by producers or distributors. However, marking out or removing this information from materials or packaging constitutes expurgation or censorship.

MPAA movie ratings, Entertainment Software Rating Board (ESRB) game ratings, and other rating services are private advisory codes and have no legal standing (*Expurgation of Library Materials*). For the library to add ratings to nonprint materials if they are not already there is unacceptable. It is also unacceptable to post a list of such ratings with a collection or to use them in circulation policies or other procedures. These uses constitute labeling, "an attempt to prejudice attitudes" (*Labels and Rating Systems*), and are forms of censorship. The application of locally generated ratings schemes intended to provide content warnings to library users is also inconsistent with the Library Bill of Rights.

The interests of young people, like those of adults, are not limited by subject, theme, or level of sophistication. Librarians have a responsibility to ensure young people's access to materials and services that reflect diversity of content and format sufficient to meet their needs.

Adopted June 28, 1989, by the ALA Council; amended June 30, 2004. [ISBN 8389-7351-5]

Available at: http://www.ala.org/ala/oif/statementspols/statementsif/ interpretations/accesschildren.htm

Access to Electronic Information, Services, and Networks:
An Interpretation of the LIBRARY BILL OF RIGHTS

Introduction

Freedom of expression is an inalienable human right and the foundation for self-government. Freedom of expression encompasses the freedom of speech and the corollary right to receive information.[1] Libraries and librarians protect and promote these rights by selecting, producing, providing access to, identifying, retrieving, organizing, providing instruction in the use of, and preserving recorded expression regardless of the format or technology.

The American Library Association expresses these basic principles of librarianship in its *Code of Ethics* and in the *Library Bill of Rights* and its Interpretations. These serve to guide librarians and library governing bodies in addressing issues of intellectual freedom that arise when the library provides access to electronic information, services, and networks.

Libraries empower users by providing access to the broadest range of information. Electronic resources, including information available via the Internet, allow libraries to fulfill this responsibility better than ever before.

Issues arising from digital generation, distribution, and retrieval of information need to be approached and regularly reviewed from a context of constitutional principles and ALA

113

policies so that fundamental and traditional tenets of librarianship are not swept away.

Electronic information flows across boundaries and barriers despite attempts by individuals, governments, and private entities to channel or control it. Even so, many people lack access or capability to use electronic information effectively.

In making decisions about how to offer access to electronic information, each library should consider its mission, goals, objectives, cooperative agreements, and the needs of the entire community it serves.

The Rights of Users

All library system and network policies, procedures, or regulations relating to electronic information and services should be scrutinized for potential violation of user rights.

User policies should be developed according to the policies and guidelines established by the American Library Association, including *Guidelines for the Development and Implementation of Policies, Regulations and Procedures Affecting Access to Library Materials, Services and Facilities.*

Users' access should not be restricted or denied for expressing or receiving constitutionally protected speech. If access is restricted or denied for behavioral or other reasons, users should be provided due process, including, but not limited to, formal notice and a means of appeal.

Information retrieved or utilized electronically is constitutionally protected unless determined otherwise by a court of law with appropriate jurisdiction. These rights extend to minors as well as adults (*Free Access to Libraries for Minors*; *Access to Resources and Services in the School Library Media Program*; *Access for Children and Young Adults to Nonprint Materials*).[2]

Libraries should use technology to enhance, not deny, access to information. Users have the right to be free of unreasonable limitations or conditions set by libraries, librarians, system administrators, vendors, network service providers, or others. Contracts, agreements, and licenses entered into by libraries on behalf of their users should not violate this right. Libraries should provide library users the training and assistance necessary to f⸍ evaluate, and use information effectively.

Users have both the right of confidentiality and the right of privacy. The library should uphold these rights by policy, procedure, and practice in accordance with *Privacy: An Interpretation of the Library Bill of Rights.*

Equity of Access

The Internet provides expanding opportunities for everyone to participate in the information society, but too many individuals face serious barriers to access. Libraries play a critical role in bridging information access gaps for these individuals. Libraries also ensure that the public can find content of interest and learn the necessary skills to use information successfully.

Electronic information, services, and networks provided directly or indirectly by the library should be equally, readily, and equitably accessible to all library users. American Library Association policies oppose the charging of user fees for the provision of information services by libraries that receive their major support from public funds (*50.3 Free Access to Information; 53.1.14 Economic Barriers to Information Access; 60.1.1 Minority Concerns Policy Objectives; 61.1 Library Services for the Poor Policy Objectives*). All libraries should develop policies concerning access to electronic information that are consistent with ALA's policy statements, including *Economic Barriers to Information Access: An Interpretation of the Library Bill of Rights, Guidelines for the Development and Implementation of Policies, Regulations and Procedures Affecting Access to Library Materials, Services and Facilities,* and *Resolution on Access to the Use of Libraries and Information by Individuals with Physical or Mental Impairment.*

Information Resources and Access

Providing connections to global information, services, and networks is not the same as selecting and purchasing materials for a library collection. Determining the accuracy or authenticity of electronic information may present special problems. Some information accessed electronically may not meet a library's selection or collection development policy. It is, therefore, left to each user to determine what is appropriate. **Parents and legal guardians who are concerned about their children's use of electronic resources should provide guidance to their own children.**

Libraries, acting within their mission and objectives, must support access to information on all subjects that serve the needs or interests of each user, regardless of the user's age or the content of the material. In order to preserve the cultural record and to prevent the loss of information, libraries may need to expand their selection or collection development policies to ensure preservation, in appropriate formats, of information obtained electronically. Libraries have an obligation to provide access to government information available in electronic format.

Libraries and librarians should not deny or limit access to electronic information because of its allegedly controversial content or because of the librarian's personal beliefs or fear of confrontation. Furthermore, libraries and librarians should not deny access to electronic information solely on the grounds that it is perceived to lack value.

Publicly funded libraries have a legal obligation to provide access to constitutionally protected information. Federal, state, county, municipal, local, or library governing bodies sometimes require the use of Internet filters or other technological measures that block access to constitutionally protected information, contrary to the *Library Bill of Rights* (ALA Policy Manual, *53.1.17, Resolution on the Use of Filtering Software in Libraries*). If a library uses a technological measure that blocks access to information, it should be set at the least restrictive level in order to minimize the blocking of constitutionally protected speech. Adults retain the right to access all constitutionally protected information and to ask for the technological measure to be disabled in a timely manner. Minors also retain the right to access constitutionally protected information and, at the minimum, have the right to ask the library or librarian to provide access to erroneously blocked information in a timely manner. Libraries and librarians have an obligation to inform users of these rights and to provide the means to exercise these rights.[3]

Electronic resources provide unprecedented opportunities to expand the scope of information available to users. Libraries and librarians should provide access to information presenting all points of view. The provision of access does not imply sponsorship or endorsement. These principles pertain to electronic resources no less than they do to the more traditional sources of information in libraries (*Diversity in Collection Development*).

[1] *Martin v. Struthers*, 319 U.S. 141 (1943); *Lamont v. Postmaster General*, 381 U.S. 301 (1965); Susan Nevelow Mart, *The Right to Receive Information* (PDF), 95 Law Library Journal 2 (2003).

[2] *Tinker v. Des Moines Independent Community School District*, 393 U.S. 503 (1969); *Board of Education, Island Trees Union Free School District No. 26 v. Pico*, 457 U.S. 853, (1982); *American Amusement Machine Association v. Teri Kendrick*, 244 F.3d 954 (7th Cir. 2001); cert.denied, 534 U.S. 994 (2001)

[3] "If some libraries do not have the capacity to unblock specific Web sites or to disable the filter or if it is shown that an adult user's election to view constitutionally protected Internet material is burdened in some other substantial way, that would be the subject for an as-applied challenge, not the facial challenge made in this case." *United States, et al. v. American Library Association* (PDF), 539 U.S. 194 (2003) (Justice Kennedy, concurring).

See Also: *Questions and Answers on Access to Electronic Information, Services and Networks: An Interpretation of the Library Bill of Rights.*

Adopted January 24, 1996, by the ALA Council; amended January 19, 2005.

[ISBN: 8389-7830-4]

http://www.ala.org/Template.cfm?Section=interpretations&Template=/ ContentManagement/ContentDisplay.cfm&ContentID=31872

Young Adults Deserve the Best:
Competencies for Librarians Serving Youth

According to a 1995 Department of Education report, public high school enrollment was expected to increase by 13% between 1997 and 2007. This increase will have a great impact on all types of libraries that serve young adults, ages 12 through 18. The need for more librarians to serve young adults is obvious. The best libraries will seize the opportunity to cultivate the increasing numbers of young adults as lifelong library partners and users.

The Young Adult Library Services Association (YALSA), a division of the American Library Association (ALA), has developed a set of competencies for librarians serving young adults. Individuals who demonstrate the knowledge and skills required by the competencies will be able to provide quality library service in collaboration with teenagers. Institutions adopting these competencies will necessarily improve overall service capacities and increase public value to their respective communities.

These competencies were developed in 1981, revised in 1998, and again in 2003 to include principles of positive youth development as they promote developmentally significant assets through excellent library service. Directors and trainers use them as a basis for staff development opportunities. They can also be used by school administrators and human resources directors to create evaluation instruments, determine staffing needs, and develop job descriptions.

The audiences for the competencies include:

- Library educators
- Graduate students
- Young adult specialists
- School library media specialists
- Generalists in public libraries
- School administrators
- Library directors
- State and regional library directors
- Human resources directors
- Non-library youth services providers
- Library grants administrators
- Youth advocacy institutions
- Youth services funding sources

Area I—Leadership and Professionalism

The librarian will be able to:

1. Develop and demonstrate leadership skills in identifying the unique needs of young adults and advocating for service excellence, including equitable funding and staffing levels relative to those provided to adults and children
2. Exhibit planning and evaluating skills in the development of a comprehensive program for and with young adults
3. Develop and demonstrate a commitment to professionalism
4. Adhere to the American Library Association Code of Ethics
5. Model and promote a non-judgmental attitude toward young adults
6. Preserve confidentiality in interactions with young adults.
7. Plan for personal and professional growth and career development through active participation in professional associations and continuing education
8. Develop and demonstrate a strong commitment to the right of young adults to have physical and intellectual access to information that is consistent with the American Library Association's Library Bill of Rights
9. Demonstrate an understanding of and a respect for diverse cultural and ethnic values
10. Encourage young adults to become lifelong library users by helping them discover what libraries offer, how to use

library resources, and how libraries can assist them in actualization of their overall growth and development

11. Develop and supervise formal youth participation, such as teen advisory groups, recruitment of teen volunteers, and opportunities for employment

12. Affirm and reinforce the role of library school training to expose new professionals to the practices and skills of serving young adults

13. Model commitment to building assets in youth in order to develop healthy, successful young adults

Area II—Knowledge of Client Group

The librarian will be able to:

1. Design and implement programs and build collections appropriate to the needs of young adults

2. Acquire and apply factual and interpretative information on youth development, developmental assets, and popular culture in planning for materials, services and programs for young adults

3. Acquire and apply knowledge of adolescent literacy, aliteracy (the choice not to read), and of types of reading problems in the development of collections and programs for young adults

4. Develop services based on sound models of youth participation and development

5. Develop programs that create community among young adults, allow for social interaction, and give young adults a sense of belonging and bonding to libraries

Area III—Communication

The librarian will be able to:

1. Form appropriate professional relationships with young adults, providing them with the assets, inputs, and resiliency factors that they need to develop into caring, competent adults

2. Demonstrate effective interpersonal relations with young adults, administrators, other professionals who work with young adults, and the community at large by:
   a. Using principles of group dynamics and group process

b. Establishing regular channels of communication (both written and oral) with each group

c. Developing partnerships with community agencies to best meet the needs of young adults

3. Be a positive advocate for young adults before library administration and the community, promoting the need to acknowledge and honor the rights of young adults to receive quality and respectful library service at all levels

4. Effectively promote the role of the library in serving young adults; that the provision of services to this group can help young adults build assets, achieve success, and in turn, create a stronger community

5. Develop effective methods of internal communication to increase awareness of young adult services

Area IV—Administration

A. Planning

The librarian will be able to:

1. Develop a strategic plan for library service with young adults based on their unique needs

2. Formulate goals, objectives, and methods of evaluation for young adult service based on determined needs

3. Design and conduct a community analysis and needs assessment

4. Apply research findings towards the development and improvement of young adult library services

5. Design, conduct, and evaluate local action research for service improvement.

6. Design activities to involve young adults in planning and decision making

7. Develop strategies for working with other libraries and learning institutions

8. Design, implement, and evaluate ongoing public relations and report programs directed toward young adults, administrators, boards, staff, other agencies serving young adults, and the community at large

9. Identify and cooperate with other youth-serving agencies in networking arrangements that will benefit young adult users

10. Develop, justify, administer, and evaluate a budget for young adult services
11. Develop physical facilities dedicated to the achievement of young adult service goals
12. Develop written policies that mandate the rights of young adults to equitable library service

B. Managing

The librarian will be able to:

1. Contribute to the orientation, training, supervision and evaluation of other staff members in implementing excellent customer service practices
2. Design, implement, and evaluate an ongoing program of professional development for all staff, to encourage and inspire continual excellence in service to young adults
3. Develop policies and procedures based upon and reflective of the needs and rights of young adults for the efficient operation of all technical functions, including acquisition, processing, circulation, collection maintenance, equipment supervision, and scheduling of young adult programs
4. Identify and seek external sources of support for young adult services
5. Monitor and disseminate professional literature pertinent to young adults, especially material impacting youth rights
6. Demonstrate the capacity to articulate relationships between young adult services and the parent institution's core goals and mission
7. Exhibit creativity and resourcefulness when identifying or defending resources to improve library service to young adults, be they human resources, material, facility, or fiscal. This may include identifying and advocating for the inclusion of interested paraprofessionals into the direct service mix
8. Document program experience and learning so as to contribute to institutional and professional memory
9. Implement mentoring methods to attract, develop, and train staff working with young adults
10. Promote awareness of young adult services strategic plan, goals, programs, and services among other library staff and in the community

11. Develop and manage services that utilize the skills, talents and resources of young adults in the school or community

Area V - Knowledge of Materials
The librarian will be able to:

1. Insure that the parent institution's materials policies and procedures support and integrate principles of excellent young adult service
2. In collaboration with young adults, formulate collection development, selection, and weeding policies for all young adult materials, as well as other materials of interest to young adults
3. Employing a broad range of selection sources, develop a collection of materials with young adults that encompasses all appropriate formats, including materials in emerging technologies, languages other than English, and at a variety of reading skill levels
4. Demonstrate a knowledge and appreciation of literature for and by young adults
5. Identify current reading, viewing, and listening interests of young adults and incorporate these findings into collection-development strategies as well as events and programs
6. Design and produce materials (such as finding aids and other formats) to expand access to collections
7. Maintain awareness of ongoing technological advances and develop a facility with electronic resources
8. Serve as a resource expert and a consultant when teachers are making the transition from textbook-centered instruction to resource-based instruction

Area VI - Access to Information
The librarian will be able to:

1. Assess the developmental needs and interests of young adults in the community in order to provide the most appropriate resources and services
2. Organize collections to maximize easy, equitable, and independent access to information by young adults
3. Use current standard methods of cataloging and classification, as well as incorporate the newest and creative means of access to information

4. Create an environment that attracts and invites young adults to use the collection
5. Develop special tools that maximize access to information not readily available, (e.g., community resources, special collections, youth-produced literature, and links to useful websites)
6. Employ promotional methods and techniques that will increase access and generate collection usage
7. Through formal and informal instruction, ensure that young adults gain the skills they need to find, evaluate, and use information effectively
8. Create an environment that guarantees equal access to buildings, resources, programs, and services for young adults
9. Develop and use effective measures to manage the Internet and other electronic resources that provide young adults with equal access
10. Develop and maintain collections that follow the best practices of merchandising

Area VII - Services

The librarian will be able to:

1. Together with young adults, design, implement, and evaluate programs and services within the framework of the strategic plan and based on the developmental needs of young adults and the public assets libraries represent
2. Utilize a variety of relevant and appropriate techniques (e.g., booktalking, discussion groups, etc.) to encourage young adult use of all types of materials
3. Provide opportunities for young adults to direct their own personal growth and development
4. Identify and plan services with young adults in non-traditional settings, such as hospitals, home-school settings, alternative education and foster care programs, and detention facilities
5. Provide librarian-assisted and independent reference service to assist young adults in finding and using information
6. Provide a variety of informational and recreational services to meet the diverse needs and interests of young adults

7. Instruct young adults in basic information gathering and research skills. These should include the skills necessary to use, evaluate, and apply electronic information sources to insure current and future information literacy.

8. Promote activities that increasingly build and strengthen information literacy skills, and develop life-long learning habits

9. Actively involve young adults in planning and implementing services and programs for their age group through advisory boards, task forces, and by less formal means (i.e., surveys, one-on-one discussion, focus groups, etc.)

10. Develop partnerships and collaborations with other organizations that serve young adults

11. Implement customer service practices that encourage and nurture positive relationships between young adults, the library, staff, and administration

---

Approved by the Young Adult Library Services Association Board of Directors, June 1981. Revised January 1998 and October 2003.

Available at: www.ala.org/ala/yalsa/professsionaldev/youngadultsdeserve.htm

Teens and Social Networking in School and Public Libraries:
A Toolkit for Librarians and Library Workers

How Online Social Networking Facilitates Learning in Schools and Libraries

What are social networking technologies? They are the types of software that enable people to connect, collaborate, and form virtual communities via the computer and/or Internet. Social networking Web sites are those that provide this opportunity to interact via interactive Web applications. Sites that allow visitors to send e-mails, post comments, build Web content, and take part in live chats are all considered to be social networking sites. These kinds of sites have come to be collectively referred to as "Web 2.0" and are considered the next generation of the Internet because they allow users to interact and participate in a way that they could not before.

Social networking technologies have many positive uses in schools and libraries. They are an ideal environment for teens to share what they are learning or to build something together online. The nature of the medium allows teens to receive feedback from librarians, teachers, peers, parents, and others. Social networking technologies create a sense of community (as do the physical library and school) and in this way are already aligned with the services and programs at the library and school.

Schools and libraries are working to integrate positive uses of social networking into their classrooms, programs, and services. By integrating social networking technologies into educational environments, teens have the opportunity to learn from adults how to be safe and smart when participating in online social networks. They also learn a valuable life skill, as these social networking technologies are tools for communication that are widely used in colleges and in the workplace. Here are a few examples of how teens are being introduced to the positive uses of social networking technologies:

- A school uses blogging software to publish its newspaper. The blog format allows for timely publication and the ability to make updates easily. This format also allows for comments from readers and easy navigation to archived stories. Publication costs are minimal (no color print costs), and there is no limit to the length of the paper, allowing for more student participation. See www.uni.uiuc.edu/gargoyle/

- An author creates a blog or MySpace account as a way to reflect on the reading and writing experience. Teens who enjoy the author's work keep up on what the author is writing and thinking through the blog. The author blog is used as a research source and as a way to communicate with the author about books, reading, and writing. See www.sparksflyup. com/weblog.php and www.myspace.com/rachel_cohn

---

Literacy and Social Networking

Social networking tools give teens meaningful ways to use and improve reading and writing skills. All social networking software requires teens to read and write. When a teen:

- Creates a profile on a social networking site
- Posts or comments on a blog
- Adds or edits content on a Wiki
- Searches for social content
- Consults peers online as a part of research,

reading and writing skills are required. This is why these technologies are referred to as the "read/write Web."

- A school librarian works with teachers to encourage student reading. As a means of getting students actively involved in their own reading, the librarian creates a Wiki where students can post their own book reviews. As they add and revise content on the Wiki, students write, read, analyze, and think critically. See www.pps-nj.us/wiki/index.php/Leeds_ Avenue_Library#Student_Reviews

- A public library creates a MySpace site as a way to connect with teens ages 14 and older in the community. The space includes quick and easy access to the library catalog and other research tools. It also includes information on programs and services at the library in which teens can take part. Teens who are not traditional library users learn about and use the library through MySpace because they are familiar and comfortable with the technology. Teens make the library one of their MySpace friends and then are reminded of the library whenever they log onto their space. See www.myspace.com/libraryloft

- A public library works with its Teen Advisory Group to set up an online del.icio.us account where teens can collect and share Web sites of interest as well as Web sites to assist each

---

Developmental Assets and Social Networking

When schools and libraries help teens use social networking tools safely and smartly, they also help teens meet their developmental assets as defined by the Search Institute (www.search-institute.org). For example, when teens:

- Learn how to use blogs, Wikis, and MySpace sites within an educational context, they learn about **boundaries and expectations**
- Are able to use social networking tools in learning, they have a **commitment to learning**
- Have the opportunity to communicate with peers, experts, authors, etc., via online social networking, they develop **social and cultural competence**
- Work with adults and peers on developing social network sites and teaching others how to use these sites, they are **empowered**
- Have a voice in the future of the school or the library, they gain a sense of **personal identity** and value
- See how librarians and teachers use social networks, they are presented with **positive role models**

other with completing homework assignments. Teens hone reading, Web searching, and critical thinking skills as they evaluate which sites to include on their del.icio.us account. See http://del.icio.us/homrteens

- A student creates a MySpace site for an author she needs to study. As she gathers information, she enters it into the writer's MySpace profile. She uses the blog function to post stories or poems she analyzes. Before long, other MySpace authors and poets (some real, some not) befriend her author. They comment on what is written and lead the student to more resources. The student has to adopt the persona of her author and imagine what the author's responses might actually be. See http://profile.myspace.com/index.cfm? fuseaction=user.viewprofile&friendid=89691844

Tips for Talking with Legislators about Social Networking

Even though librarians are respected members of the community, competition for attention and time of elected officials is great, as is competition for funding. It is important that librarians reach out to elected officials and educate them about the needs of libraries and library patrons.

1. Communicate via phone, fax, or in person. If you are hoping to meet with a legislator in person, set up an appointment in advance. (By the way, do not be disappointed if you end up communicating with someone from the legislator's staff.)

2. Be polite, respectful, professional, and friendly.

3. Introduce yourself, identify your job title, and state your purpose.

4. Stick to the point: communicate ONE message—the benefits of social networking for teens.

5. Use specific examples from your own work with teens to illustrate your point. If you are meeting the legislator in person, you might even be able to take a well-spoken teen and/or parent with you who can talk about the benefits of social networking.

6. Ask for action. For example, ask the legislator to vote against any legislation that attempts to restrict or ban social

Before You Visit

Do your homework. Find out what legislation is before the state senate and be aware of what it says and where the person you are going to talk to stands on the issues related to social networking.

Gather personal stories relating to the issue from your teen library patrons and their parents to share with the legislator.

Visit or contact your legislator as soon as you hear about pending legislation.

You can find out about state and national social networking legislation at http://wikis.ala.org/iwa, or go to www.ala.org and click on "Take Action."

networking sites in libraries. Or ask the legislator to support any legislation that supports social networking and Internet access, like the E-rate.

7. Offer to provide additional information about social networking. Take such materials with you if you are meeting the legislator in person.

8. Listen carefully and courteously.

9. Invite the legislator to visit your library. Provide a calendar of events.

10. Remember to say "thank you."

Educating the Community about Online Social Networking

There are many examples in the media of how social networking has played a dangerous role in teen lives. However, positive examples of how this technology supports teen literacy skills and developmental growth are not so readily found. For that reason, librarians should play an active role in educating parents, teachers, and other members of the community about the positive benefits of social networking in teen lives. The following examples of how you can educate your community provide a starting point. When planning these events, be sure to enlist your Teen Advisory Group (TAG) to help you plan and implement the workshops.

• Create and distribute brochures and post information online about what your library is already doing to ensure that children and teens are safe online. Include information about

Internet filters and Internet Acceptable Use Policies that your library has.

- Invite parents and educators to a workshop where they can learn about MySpace, Facebook and other social networking tools. In the workshop, have librarians and teachers discuss how MySpace is being used in the classroom and library. Have law enforcement officials talk about how to help teens stay safe while participating in social networking online. Have teens with well-designed MySpace pages demonstrate the positive ways they use social networking tools.

- Host Do-It-Yourself Days for adults to learn how to use social networking sites and tools successfully. After an introduction about what social networking is and why it is an important part of a teen's life, teens from your TAG could work with adults on using the tools in a way that enhances their own lives. Teens might show adults how to set up: a blog that showcases a hobby or special interest; a MySpace page to keep in touch with friends; a Flickr account so they can share family photos; an IM account to conduct live chats with family members overseas; and more.

- Create an online demonstration or class that gives adults a chance to test out and discuss social networking technologies at their leisure and in a somewhat anonymous setting. Make the demo available from your library's Web site. Use your TAG group to help develop the demo.

Sites That May Be Affected by Social Networking Legislation

There are many sites currently used by adults and teens that will be blocked in schools and libraries if legislation prohibiting access to social networking sites passes. These include:

- Photo-sharing sites like Flickr.com, which patrons use to share photos with family members who are far away
- Health-related sites like PsychCentral.com, which allow users to get important medical questions answered during live chat sessions
- Educational sites like LegalGuru.com, which allows users to get free legal advice
- Library reference sites, where patrons can get questions answered via instant messaging that use AOL, Yahoo!, or other commerical services.

- Host a community debate about local, state, or national legislation that seeks to regulate social networking Web sites. Invite local experts on both sides of the issue to participate in the debate. Provide handouts and background information for attendees.
- Use social networking technologies as an access point for your library's services. Create a MySpace page as a place for adults and teens to learn out about programs and materials. Set up a blog where adults and teens read about what is going on in the library and can add comments about programs, materials, and so on. Develop a booklist Wiki where adults and teens can add titles of books on specific themes.
- Inform—perhaps via a podcast—educators, parents, and community members about how social networking tools allow for schools and libraries to integrate technology in meaningful ways with and for teens at low (or no) cost. Information could include overviews of the technologies, interviews with teens about their use of technology, interviews with experts in technology and teen development who discuss how the technologies support teen growth and literacy development, and so on.
- Create and distribute an information sheet for adults about the positive aspects of social networking as well as Internet safety tips, and include annotated lists of resources. You can also post the information on your library's Web site, blog, Wiki, or MySpace pages.
- Sponsor a scholarly presentation, or series of presentations, for local educators and concerned adults by experts in the field of developmental assets, teen print literacies in the world of technology, and social networking. Ask speakers to focus directly on how social networking technologies have positive benefits for teens.
- Host an evening that focuses on how social networking is being used in higher education and business. Invite faculty from a local college or university to talk to parents and teens about how they use social networking technologies with students to facilitate the teaching and learning process. Invite business leaders to talk about what social networking technologies their employees must know how to use in order to be successful in their jobs.

Social Networking Defined

The following definitions of social networking tools should help you explain to your community what it is all about.

Blog: A Web page where you can write journal entries, reviews, articles, and more. Blog authors can allow readers to post their own comments. No Web design knowledge is needed to create a blog.

Podcasts: Audio files that are available for download. They are usually available via subscription so that you can automatically download it to a computer or MP3 player (like an iPod).

RSS: A way for subscribers to automatically receive information from blogs, online newspapers, and podcasts.

Social Networking: In the online world, this refers to the ability to connect with people through Web sites and other technologies, like discussion boards.

Tagging: Refers to the ability to add subject-headings to content in order to organize information in a meaningful way and to connect others that tag similar content in the same way.

Wiki: A collaborative space for developing Web content. No Web design knowledge is needed to create a Wiki.

Educating Teens about Online Social Networking

You can help teens use social networking technologies successfully and safely by sponsoring programs and services that focus on these technologies. The following examples are available to help you get started. Show these examples to your Teen Advisory Group (TAG) and see which one(s) they feel are important to offer in your community. Have your TAG help plan and carry out the event(s). Remember that social networking sites often have minimum age requirements and be sure to honor those.

- Offer a class to teach teens how to use Blogger.com. As teens set up their blog, you can facilitate a discussion about Internet safety issues, the importance of guarding against identity theft, online etiquette, etc.

- Host Do-It-Yourself Days for teens where they learn about a variety of social networking technologies. You might have a day for photo-sharing technologies, another day for bookmarking sites, another day for friend building, and so

Social Networking Sites

del.icio.us
www.del.icio.us
Combines bookmarking and tagging. Allows users to network with others in order to keep track of what is being bookmarked by those with similar interests.

Flickr
www.flickr.com
A photo-sharing site that allows users to tag images.

MySpace
www.myspace.com
Users build their space on the Web and then invite others to be their friends.

Technorati
www.technorati.com/
A searchable database of blogs that gives bloggers the ability to tag content for easy access by others.

Wikipedia
www.wikipedia.org/
A Wiki encyclopedia that gives anyone the ability to add to and change entries.

on. During each of the sessions you can talk with teens about how to make decisions about safe use of these technologies.

- Work with teens to produce podcasts on topics of interest. They might review media and books, talk about what's going on in the community, book talk, etc. As a part of the podcast process, have teens write outlines of the content they want to cover and talk with them about whom they want to make the podcast available to.

- Together with teens, create a library books and media Wiki as a means for recommending resources to library patrons. Train teens on how to update the content of the Wiki and talk about how to evaluate the quality of information in Wikis and other types of resources.

- Take photos at the library and have teens upload and tag them on Flickr or another photo-sharing site. As a part of the uploading and tagging process, discuss safety and privacy concerns with teens and decide whether or not the photos should be private or public. As they tag the photos, ask them to consider what the best ways are to describe content in order for friends or the public (if the photos are made public) to find them.

- Work with teens to create a Wiki, podcast, or Web page about Internet safety aimed at children. Post the completed resource on your library's Web site.

- Have teens create del.icio.us accounts for collecting resources they can use in school research. The teens can network with classmates and peers in del.icio.us in order to learn about resources their peers have uncovered that support learning on a particular topic. Use del.icio.us networking as a jumping off point for a discussion of evaluating information quality.

- Use Flickr as a platform for creative writing exercises with teens. Upload your own or teens' photos to Flickr and then have teens tell a story with the photos through captions that they add.

- Invite a technology expert in to talk with teens about how social networking tools work.

Some Authors Using MySpace

Rachel Cohn	**Sara Dessen**	**Brian Sloan**
www.myspace.com/ rachel_cohn	www.myspace.com/ sarahdessen	www.myspace.com/ bmsloan
John Green	**Melissa de la Cruz**	**Ned Vizzini**
www.myspace.com/ greatperhaps	www.myspace.com/ melissadelacruz	www.myspace.com/ nedvissini
Cecil Castellucci	**David Levithan**	
www.myspace.com/ cecilseaskull	www.myspace.com/ davidlevithan	

Some Libraries Using MySpace

Bethpage Public Library	**Lansing Library**
www.myspace.com/ bethpagepubliclibrary	www.myspace.com/ lansingpubliclibrary
Hennepin County Library	**Library Loft**
www.myspace.com/ hennepincountylibrary	www.myspace.com/ libraryloft

Tools to Use to Get Started with Social Networking

Blogging
Set up blogs with a free service like Blogger.com (www.blogger.com), or download and customize software with Word Press (www.wordpress.com/)

Wikis
Setup a Wiki with a free tool such as WikiSpaces (www.wikispaces.com), or download and customize software with MediaWiki (www.mediawiki.org/)

- Give teens the chance to connect with favorite authors, artists, musicians, and so on via MySpace and personal blogs. Teens can search for the spaces and blogs using common searching tools and then comment on the blogs and MySpace pages of those they connect with.

- Build a library MySpace page with teens. Have teens meet to plan the space, including what it should look like and include. Work with them to build the site and develop guidelines for blogging, commenting, and making friends on the site. As a part of this project, talk with teens about how to decide whether or not to accept those who want to befriend them on MySpace. Add value to your MySpace presence through links to online safety and library resources. Make it possible for teens to add your catalog search on their MySpace accounts.

Additional Resources about Online Social Networking and Libraries

For Librarians and Educators

Discuss DOPA
http://www.saveyourspace.org/
Articles and information on social networking legislation and a Save Your Space petition.

DOPA Watch
www.andycarvin.com/dopa.html
The latest news on legislation related to social networking in schools and libraries.

The Gateway to Educational Materials
www.thegateway.org

Search with the term "Internet safety" to locate resources and lesson plans about Internet safety.

Goodstein, Anastasia. 2007. *Totally Wired: What Teens Are Really Doing Online.* New York: St. Martin's.

Interactive Web Applications Wiki
http://wikis.ala.org/iwa
A Wiki that tracks recent social networking legislation and provides resources for library workers.

The Internet & Teens: Social Networking Safety, by Bill Erbes
www.opal-online.org/archivelis.htm
An online tutorial for librarians about teaching teens responsible use of social networking sites.

Internet Safety
www.ala.org/ala/washoff/WOissues/techinttele/internetsafety/internetsafety.htm
Multiple resources are accessible from this page, including an FAQ on libraries, children, and the Internet, and a toolkit about libraries and the Internet from ALA's Office of Intellectual Freedom.

Librarians' Index to the Internet
Two annotated lists of Web sites are particularly useful:
• Social Networks: www.lii.org/pub/subtopic/4679
• Internet Safety: www.lii.org/pub/subtopic/948

Prensky, Marc. 2005. "Adopt and Adapt." *Edutopia.*
Available at: www.edutopia.org/magazine/ed1article.php?id=Art_1423&issue=dec_05#
Prensky covers why it's important for schools to integrate new technologies into the classroom.

Prensky, Marc. 2005. "Engage Me or Enrage Me: What Today's Learners Demand." *Educause* (September/October).
Available at: www.educause.edu/ir/library/pdf/erm0553.pdf
A look at why using technologies that are of interest to students improves learning.

Richardson, Will. 2006. *Blogs, Wikis, Podcasts and Other Powerful Web Tools for the Classroom.* Thousand Oaks, CA: Corwin Press.
Richardson explains how and why social networking can be used in the library and classroom.

Teen Content Creators and Consumers
www.pewinternet.org/PPF/r/166/report_display.asp

This Pew *Internet in American Life* report illustrates how and why teens use technology to communicate and create information.

30 Positive Uses of Social Networking
www.leonline.com/yalsa/positive_uses.pdf
Provides ideas from librarians about how social networking can be integrated into schools and libraries successfully. Use these ideas to educate your colleagues, peers, and government officials about how social networking plays a positive role in teen lives.

Weblogg-ed
www.weblogg-ed.com/
Educator and technology specialist Will Richardson discusses how and why new technologies should be integrated into the classroom and library on his frequently updated blog.

For Teens

Social Networking Sites: Safety Tips for Tweens and Teens
www.ftc.gov/bcp/edu/pubs/consumer/tech/tec14.htm
A short and useful list of reminders for staying safe on social networking sites (and online in general). Includes a list of resources for finding out more.

For Parents and Caregivers

Farnham, Kevin, and Dale Farnham. 2006. MySpace Safety: 51 Tips for Parents and Teens.
www.howtoprimers.com
Two parents discuss how to ensure teens are safe when using social networking.

Willard, Nancy. 2007. *Cyber-Safe Kids, Cyber-Savvy Teens: Helping Young People Learn to Use the Internet Safely and Responsibly.* San Francisco, CA: Jossey-Bass.

Wired Safety.org: Blog Sites, Profile Sites, Diary Sites or Social Networking Sites.
www.wiredsafety.org/internet101/blogs.html
Information on what parents need to do in order to help their children stay safe when using social networking technologies.

For Everyone

iSAFE
www.isafe.org

Provides resources about Internet safety. There is a different section of the site each for parents, educators, kids and teens, and law enforcement. There are free online tutorials for young people and adults, as well as printable newsletters and other resources.

NetSmartz

www.netsmartz.org.

Provides resources about Internet safety. There is a different section of the site each for parents, educators, kids, teens, press, and law enforcement. Maintained by the National Center for Missing & Exploited Children.

Index

About the Author

RoseMary Honnold is the Young Adult Services Coordinator at Coshocton Public Library in Coshocton, Ohio.

In addition to being an amateur painter, Honnold finds that reading, writing, programming, presenting workshops, and working full-time keeps her busy, while her grandchildren, Caleb, Evelyn, and Wesley, keep her priorities in order.

She is the author of *Serving Seniors: A How-To-Do-It Manual for Librarians*, *101+ Teen Programs that Work*, *More Teen Programs that Work*, and *The Teen Reader's Advisor*, all published by Neal-Schuman.